MATHEMATICS INSTRUCTION: GOALS, TASKS AND ACTIVITIES

Yearbook 2018

Association of Mathematics Educators

Related Titles

Empowering Mathematics Learners, Yearbook 2017
edited by Berinderjeet Kaur and Ngan Hoe Lee
ISBN: 978-981-3224-21-6

Developing 21st Century Competencies in the Mathematics Classroom, Yearbook 2016
edited by Pee Choon Toh and Berinderjeet Kaur
ISBN: 978-981-3143-60-9

Effective Mathematics Lessons through an Eclectic Singapore Approach, Yearbook 2015
by Khoon Yoong Wong
ISBN: 978-981-4696-41-8

Learning Experiences to Promote Mathematics Learning, Yearbook 2014
edited by Pee Choon Toh, Tin Lam Toh and Berinderjeet Kaur
ISBN: 978-981-4612-90-6

Nurturing Reflective Learners in Mathematics, Yearbook 2013
edited by Berinderjeet Kaur
ISBN: 978-981-4472-74-6

Reasoning, Communication and Connections in Mathematics, Yearbook 2012
edited by Berinderjeet Kaur and Tin Lam Toh
ISBN: 978-981-4405-41-6

Assessment in the Mathematics Classroom, Yearbook 2011
edited by Berinderjeet Kaur and Khoon Yoong Wong
ISBN: 978-981-4360-97-5

Mathematical Applications and Modelling, Yearbook 2010
edited by Berinderjeet Kaur, Jaguthsing Dindyal
ISBN: 978-981-4313-33-9

Mathematical Problem Solving, Yearbook 2009
edited by Berinderjeet Kaur, Ban Har Yeap, Manu Kapur
ISBN: 978-981-4277-20-4

MATHEMATICS INSTRUCTION: GOALS, TASKS AND ACTIVITIES

Yearbook 2018
Association of Mathematics Educators

editors

Pee Choon Toh
Boon Liang Chua

Nanyang Technological University, Singapore

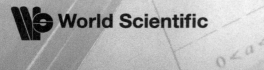

Published by

World Scientific Publishing Co. Pte. Ltd.
5 Toh Tuck Link, Singapore 596224
USA office: 27 Warren Street, Suite 401-402, Hackensack, NJ 07601
UK office: 57 Shelton Street, Covent Garden, London WC2H 9HE

British Library Cataloguing-in-Publication Data
A catalogue record for this book is available from the British Library.

MATHEMATICS INSTRUCTION: GOALS, TASKS AND ACTIVITIES
Yearbook 2018, Association of Mathematics Educators

ISBN 978-981-3271-66-1

For any available supplementary material, please visit
http://www.worldscientific.com/worldscibooks/10.1142/11027#t=suppl

Printed in Singapore

Contents

Chapter 1

Tasks and Activities in the Mathematics Classroom

Boon Liang CHUA   Pee Choon TOH

This introductory chapter provides an overview of the chapters in the book. The chapters are classified into three parts. The first part considers different sets of design principles and affordances of mathematical tasks. The second part is a showcase of tasks and activities developed by various mathematics educators towards the goal of helping students learn effectively. The third part emphasises the important role that 'A' level mathematics teachers play in implementing mathematical tasks.

1 Mathematics Instruction: Goals, Tasks and Activities

This yearbook of the Association of Mathematics Educators in Singapore invites mathematics teachers to reflect and rethink their beliefs about mathematics teaching and learning in the 21st century to improve mathematics instruction in the classroom. Like the previous yearbooks such as *Developing 21st Century Competencies in the Mathematics Classroom* (Toh & Kaur, 2016), and *Empowering Mathematics Learners* (Kaur & Lee, 2017), the theme of this present yearbook is shaped not only by the school mathematics curriculum developed by the Ministry of Education and the needs of mathematics teachers in Singapore schools, but also the current international trend in mathematics education.

   In recent years, there is a growing interest amongst mathematics educators in task design. The growth of attention to task design is not surprising at all when viewed through the lenses of practical, cognitive and cultural perspectives. From the practical perspective, mathematical tasks are the bedrock of mathematics lessons in many countries (Watson & Ohtani, 2015). Mathematics classroom instruction is normally organised and delivered through the activities in mathematical tasks developed by mathematics teachers or found in curriculum materials. In other words, mathematics instructions are information guiding students how to carry out the activities in the mathematical task. They may take different forms, but typically as a series of easy-to-follow steps and guiding questions that prompt students to explore and discover the mathematical concepts embodied in the task. To examine classroom instructions, it is common to look at and analyse the amount of lesson time that students spend on doing the task (Shimizu, Kaur, Huang, & Clarke, 2010). For instance, the Year 8 students in all seven participating countries in the TIMSS 1999 Video Study were found to have spent at least 80% of their time in mathematics lessons working on mathematics tasks (Hiebert et al., 2003). Watson and Ohtani (2015) also point out, from the cognitive viewpoint, that the goal and content of a mathematical task are important and can have significant effect on students' learning. In a similar way, they highlight the cultural aspect of mathematical tasks when they remark that mathematical tasks can shape students' learning experiences of mathematics and their understanding of the nature of mathematical activity.

2 Mathematical Tasks vs Mathematical Activities

Mathematical tasks are crucial vehicles in the classroom for enhancing students' mathematical thinking and reasoning (Stein, Grover, & Henningsen, 1996; Watson & Ohtani, 2015). The phrase "mathematical task" is often used alongside "mathematical activity" by many authors in the mathematics education literature, but the distinction between them is rarely noticed and discussed by the authors.

According to Watson and Sullivan (2008), mathematical tasks refer to the questions, situations and instructions that are accessible to students in the mathematics lessons. So a task could range from a mathematics textbook exercise, or an examination question, to an exploration using the guided-discovery approach. Tasks such as these, regardless of their cognitive demands, embody mathematical ideas and trigger activities which then offer students opportunities to encounter mathematical concepts, skills and processes (Mason & Johnston-Wilder, 2006; Watson & Ohtani, 2015). Consider the factorisation task involving the *Multiplication Frame* method in Chua (2017). The learning goal of the task for students is to master the technique of factorising quadratic expressions of the form $ax^2 + bx + c$ using the *Multiplication Frame* method. However, the activities within the task that students have to engage with include examining how the terms are positioned in the frame and establishing the relationship between the pair of quadratic term and the constant, and the pair of linear terms in the frame. Hence, mathematical tasks are different from mathematical activities. In Chapter 2 of this book, *From Task to Activity: Noticing Affordances, Design, and Orchestration*, Choy draws the same distinction between mathematical tasks and mathematical activities, and argues why such a distinction is crucial.

In the following sections, 13 peer-reviewed chapters resulting from the keynote lectures and workshops from the Mathematics Teachers Conference 2017 will be introduced. These chapters are classified broadly into three parts. In the first part are two chapters who offer sets of principles for designing mathematical tasks so as to provide meaningful learning experiences to students.

The second part of the book comprises nine chapters that showcase mathematical tasks and activities developed by educators to facilitate effective mathematics learning in the classrooms. The content of these chapters varies widely in the choice of teaching approaches, ranging from flipped classroom and the use of comics to mathematical modelling and problem posing.

In the third part that consists of two chapters, the role of 'A' level mathematics teachers in task design and implementation is discussed. The authors call for mathematics teachers to deepen their understanding of curriculum policy and mathematics content so that they can enrich the learning experiences of their Grades 11 and 12 students. The following three sections provide summaries of each chapter, specifically highlighting how the chapter addresses the theme of the book, *Mathematics Instruction: Goals, Tasks and Activities*.

3 Task Design Principles and Affordances of Mathematical Tasks

In the second chapter *From Task to Activity: Noticing Affordances, Design, and Orchestration*, Choy observes that very few teachers perceive task and activity as totally different things and illustrates the difference between the two using an example of a task on investigating the graph of a trigonometric function using graphing software. He also points out how important it is for mathematics teachers to notice the mathematics embedded in a task and to think about how to use the task to orchestrate meaningful learning experiences for their students. He then introduces the MAD (Mathematics, Activity and Documentation) framework and follows by demonstrating how to orchestrate discussions in the mathematics lessons using the five practices developed by Stein, Engle, Smith, and Hughes (2008). Additionally, Choy draws a vignette to show how a typical examination question can be used beyond the purpose of honing procedural skills to engage students in mathematical thinking about the concept.

In Chapter 3, *Affordances of Typical Problems*, Dindyal offers a description of typical problems as mathematical tasks that are readily available and are often used to develop procedural skills. He holds the view that the affordances of typical problems for developing conceptual fluency can only be realised when mathematics teachers are able to modify them in various ways. He then suggests eight strategies, with examples, to illustrate how typical problems can be modified and then used productively in mathematics lessons.

4 Mathematical Tasks and Activities for Effective Learning

In *Mathematical Tasks Enacted by Two Competent Teachers to Facilitate the Learning of Vectors by Grade Ten Students* (Chapter 4), Kaur, Wong, and Chew describe how two competent teachers, each with 20 years of teaching experience, enacted mathematical tasks to facilitate the learning of vectors. The teachers used similar types of tasks to achieve the goals of their lessons. But their tasks differed in context and cognitive demands due to different students' interest and ability. The teachers also hold different beliefs in how students should learn. One believes in mastery by practice after the mathematical concept is properly understood whereas the other believes in both developing understanding and building skills for the topic. Kaur et al. emphasise the important role that mathematics teachers play when they enact the mathematical tasks so as to support meaningful connection amongst concepts, procedures and contexts, and to provide opportunities for students to engage in mathematical reasoning and problem solving.

Chapter 5, *Use of Comics and Its Adaptation in the Mathematics Classroom*, by Toh, Chan, Cheng, Lim, and Lim describes the case of two mathematics teachers from a secondary school co-teaching the topic of Percentages using comics. The teachers were mindful of the instructional goals and adapted the tasks in the comics teaching package developed by Toh et al. accordingly to enhance students' conceptual fluency as well as procedural fluency.

In *Designing and Implementing Scientific Calculator Tasks and Activities* (Chapter 6), Kissane considers the educational potential of scientific calculators in mathematics learning, including offering different *representations* of the same result, allowing *computation*, encouraging *exploration*, and seeking *affirmation*. He points out that for calculator tasks to achieve maximum educational value, providing teachers with adequate guidance on the implementation of the tasks is crucial. The teacher guide can provide information such as the answers to the tasks, the nature and purpose of the task, and suggested classroom organisation and time needed to complete the task. Examples of calculator tasks, with analysis of some of them, are provided to show mathematics teachers how worthwhile tasks can be designed.

In *Engaging the Hearts of Mathematics Learners* (Chapter 7), Joseph Yeo B. W. introduces the *LOVE Mathematics* framework (*L*inking *O*pportunities in a *V*ariety of *E*xperiences to the learning of mathematics) to illustrate how students can be engaged in mathematics lessons through three principles: namely, *variety, opportunity*, and *linkage*. Mathematics activities incorporating the use of amusing mathematics videos, catchy mathematics song, witty mathematics comics, puzzles and games are described to demonstrate how mathematics lessons can become more engaging.

In Chapter 8, *Developing Interaction Toward the Goal of the Lesson in a Primary Mathematics Classroom*, Hino maintains that the goal of the lesson that a mathematics teacher has in mind shapes the mathematics instruction in the classroom. Drawing evidence from classroom interactions in two Grade 5 mathematics lessons on comparing fractions taught by an experienced teacher, she illustrates how being clear about the goal of the lesson helped the teacher not only engage his students in meaningful interactions but also manage various students' responses. For instance, the mathematics teacher, other than guiding his students to develop conceptual understanding through questioning, also paid particular attention to their choice of methods as well as use of mathematical language and notations.

Chapter 9 by Cheng, Ng, Tan, and Ng, *Designing and Implementing Activities in the Flipped Classroom in the Singapore Primary Mathematics Classroom*, describes a teaching model involving the flipped classroom approach that was used by a group of Primary 5 mathematics teachers to teach the topic of triangles. The teachers' involvement in the development of mathematical tasks for the teaching package has benefitted them greatly. Cheng et al. describe both the benefits gained and challenges encountered by the teachers in the chapter.

The next chapter, *Designing Mathematical Modelling Activities for the Primary Mathematics Classroom* (Chapter 10), by Chan, Vapumarican, and Liu, offers those who are new to mathematical modelling a brief description of the Model-Eliciting Activities perspective, which involves complex, open, and non-routine modelling tasks situated in a real-world context to allow students to exercise both informal and formal mathematical knowledge interactively. Chan et al. present the design principles and exemplify the principles through two examples, one on recommending a suitable phone plan and the other on modelling the spread of mosquito borne diseases.

Yeo K. K. Joseph offers two strategies to modify and extend closed textbook exercises typically used in primary schools into short open-ended tasks in the chapter *Extending Textbook Exercises into Short Open-Ended Tasks for Primary Mathematics Classroom Instruction* (Chapter 11). He also discusses the benefits of implementing open-ended tasks and the implications for teaching and learning from using such tasks.

The last chapter in the second part of the book, *Integrating Problem Posing into Mathematical Problem Solving: An Experimental Study* (Chapter 12) by Jiang and Chua, explores the use of the what-if-not strategy to encourage students to pose problems in an experimental study involving 56 Grade 7 students in Macao. The study, conducted over four lessons on the topic of solving simultaneous linear equations in two variables, involved a pre-test and a post-test. The results show that students exposed to the what-if-not strategy performed better in both problem solving and problem posing items.

5 Role of 'A' Level Mathematics Teachers

The two chapters in this third part of the book underscore the role of teachers in selecting and designing mathematics tasks for use in classroom instruction. In Chapter 13, *A Vicennial Walk Through 'A' Level Mathematics in Singapore: Reflecting on the Curriculum Leadership Role of the JC Mathematics Teacher*, Ho and Ratnam-Lim call for teachers to be curriculum leaders. By curriculum leaders, they expect teachers and heads of department to take on the responsibility of understanding the purpose for changes in the syllabus, the shifts in educational orientations, and how these are then translated into the scope and sequence of teaching and learning experiences and assessments. By assuming the role of an active curriculum leader, they believe that teachers are then able to formulate goals for the lessons, design mathematical tasks and implement classroom activities to bring about a more enriching learning experience for 'A' level students.

Chapter 14 by Yap, *Probability: Theory and Teaching*, is the final chapter of this yearbook. Yap, a statistician, was invited to lecture on a mathematical topic that many teachers find difficult to teach. This chapter explains briefly Kolmogorov's axioms and explores three different meanings of probability, paying particular attention to the frequency interpretation. Yap provides a list of ten problems and their solutions, with the aim of strengthening the appreciation of the frequency approach in probability.

6 Concluding Remarks

Mathematical tasks are important vehicles for classroom instruction to bring about positive impact on student learning. The relationship between mathematical tasks, teaching and learning needs to be managed carefully by the teachers. Otherwise, gaps may appear between what the teacher on one side intends and what students on the other side perceive.

The chapters in this yearbook provide readers and specifically classroom teachers with a cache of resources to help them implement mathematical tasks in lessons. Inside the cache, readers will find the different task design principles, different types of mathematical tasks used in classroom instruction and teaching approaches to implement the tasks. Readers are encouraged to read the chapters carefully, try some of the ideas in their classrooms, and convince themselves that these ideas offer a means to provide enriching learning experiences to students.

References

Chua, B. L. (2017). Empowering learning in an Algebra class: The case of expansion and factorization. In B. Kaur, & N. H. Lee (Eds.), *Empowering mathematics Learners* (pp. 9-30). Singapore: World Scientific.

Hiebert, J., Gallimore, R., Garnier, H., Givvin, K. B., Hollingsworth, H., Jacobs, J., ... Stigler, J. W. (2003). *Teaching mathematics in seven countries: Results from the TIMSS 1999 video study*. Washington, DC: NCES.

Kaur, B., & Lee, N. H. (2017). *Empowering mathematics learners*. Singapore: World Scientific.

Mason, J., & Johnston-Wilder, S. (2006). *Designing and using mathematical tasks*. United Kingdom: Tarquin Publications.

Shimizu, Y., Kaur, B., Huang, R., & Clarke, D. (2010). The role of mathematical tasks in different cultures. In Y. Shimizu, B. Kaur, R. Huang, & D. Clarke (Eds), *Mathematical tasks in classrooms around the world* (pp. 1-14). Rotterdam: Sense Publishers.

Stein, M. K., Engle, R. A., Smith, M. S., & Hughes, E. K. (2008). Orchestrating productive mathematical discussions: Five practices for helping teachers move beyond show and tell. *Mathematical Thinking and Learning, 10*(4), 313-340.

Stein, M. K., Grover, B. W., & Henningsen, M. (1996). Building student capacity for mathematical thinking and reasoning: An analysis of mathematical tasks used in reform classrooms. *American Educational Research Journal, 33*(2), 455-488.

Toh, P. C., & Kaur, B. (2016). *Developing 21st century competencies in the mathematics classroom*. Singapore: World Scientific.

Watson, A., & Ohtani, M. (2015). Themes and issues in mathematics education concerning task design: Editorial introduction. In A. Watson, & M. Ohtani (Eds), *Task design in mathematics education* (pp. 3-18). Heidelberg: Springer.

Watson, A., & Sullivan, P. (2008). Teachers learning about tasks and lessons. In D. Tirosh, & T. Wood (Eds.), *Tools and resources in mathematics teacher education* (pp. 109-135). Rotterdam: Sense Publishers.

Chapter 2

From Task to Activity:
Noticing Affordances, Design, and Orchestration

CHOY Ban Heng

How teachers perceive and use tasks in mathematics classrooms can potentially enhance the learning experiences of our students. Many teachers perceive task and activity to be the same thing. Are they really the same? What we do know is that teachers need to notice the mathematics embedded in a task, and think about how they use tasks to orchestrate mathematically meaningful learning experiences for their students. However, doing this work of orchestrating learning experiences is challenging for teachers, and so, how can we support teachers to use tasks effectively? In this chapter, I will argue why a distinction between task and activity is critical, provide three principles for designing or selecting tasks, and suggest how teachers can orchestrate productive discussions when using tasks for teaching. In addition, I will highlight, through a vignette drawn from an ongoing study, how teachers can think about using typical problems—examination or textbook type questions—beyond their usual purpose of honing procedural skills to provide more engaging learning experiences for students.

1 Orchestrating Learning Experiences with Tasks

Learning experiences are included as part of the current mathematics syllabus to provide opportunities for students to engage in mathematical processes, and to "influence the ways teachers teach and students learn" (Ministry of Education, 2013, p. 20), so that the key objectives of mathematics education in Singapore can be achieved. They are stated in the form "students should have opportunities…" to signal the type of activities expected for each topic. Although the intentions and even the descriptions of learning experiences are clearly spelt out, teachers are given the autonomy to design, select, and adapt the suggested tasks to provide these experiences for students. However, as Tyler (1949) highlights, different students may experience the same task differently even though the task is set up in the same way. This, according to Tyler (1949), presents the challenge of setting up tasks to initiate learning activities. The crux lies in teachers being aware of the kind of activities, which students engage in when working on a task.

1.1 *Task or Activity?*

Tasks and activities are sometimes taken to mean the same thing, but Mason and Johnston-Wilder (2006) distinguish them by highlighting that "the purpose of a task is to initiate activity by learners" (p. 5). In other words, a task is a "prompt" or set of instructions for students' work (Sullivan, Clarke, & Clarke, 2013, p. 13), whereas an activity refers to the kind of mathematical processes such as mathematical reasoning students engage in when they work on the task. This distinction is important because students may work on the task but do not engage in the activity. In other words, they may not have acted upon the physical, symbolic, or mental objects, in order to become sensitised to important features in the concept even though they might have been working at the task (Mason & Johnston-Wilder, 2006).

For example, the curriculum documents suggest the use of graphing software to display the graphs of trigonometric functions and discuss their behaviours, and investigate how a graph, e.g., $y = a \sin bx + c$ changes when a, b, or c varies (see Figure 1). This task is common and many teachers, including me, have used it when teaching trigonometric functions. However, as I have observed, students may not focus on examining how the graph of the function changes with the values of a.

Instead, they may be more interested in 'playing with the sliders' rather than thinking about the relationship between the parameters and the shape of the graph.

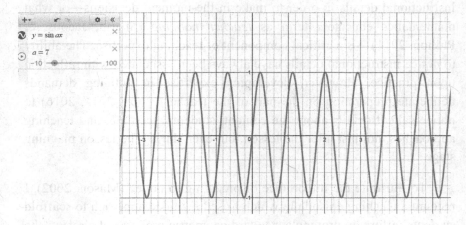

Figure 1. Using graphing software to explore trigonometric functions

This example highlights one of the key challenges of using tasks in the mathematics classroom: the task is different from the activity. Consequently, while the design of a task is crucial for facilitating and encouraging students' development of the mathematical processes, it is equally important to take note of how students engage with the task. In particular, teachers should pay attention to the kind of mathematical activity intended in the task, and start thinking about how they can encourage students to engage in the intended activities during task implementation.

1.2 *Noticing task design and orchestrating discussions*

Thinking about how teachers can facilitate the mathematical activity as intended by the task is challenging, and this requires "an unnatural orientation" and "a simultaneous, unusual attention" to both the learners and the content to be taught (Ball & Forzani, 2009, p. 499). It thus involves a shift of attention (Mason, 2010), on the part of the teacher, to listen to and examine students' thinking carefully in order to identify the key mathematical ideas and address corresponding students' errors (Ball & Forzani, 2009; Schifter, 2001). Attending to and making sense of relevant instructional details in order to make in-the-moment decisions—or what mathematics educators term as *teacher noticing* (Sherin, Jacobs, & Philipp, 2011)—is a critical component of teaching expertise. The ability to notice instructional details have always been associated with reflection upon practice. However, developing expertise in noticing demands deliberate preparation on the part of the teacher (Choy, 2014, 2016) to notice salient details about the content, students' learning, and teaching approaches, not just after the lesson, but even during the lesson planning stage.

To illustrate the importance of preparing to notice (Mason, 2002), I recount a teaching episode in which I used a guided approach to scaffold students' efforts in proving a standard geometric property of circles—the angle at the centre is twice the angle at the circumference subtended by the same arc. I considered two cases (see Figure 2) in my proving task.

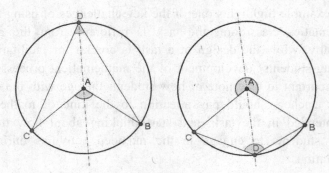

Figure 2. The two cases used in my proof

The students worked on the task and engaged with the mathematical processes of reasoning. Everything went well until one of the students, John (a pseudonym), raised his hands and asked me to look at a case, which he had drawn (see Figure 3). I realised, at that point, that John's case is different from the two cases I had used in my task. Although I could use another property—angles in the same segment are equal—to prove the result, John asked whether there was a more elementary proof that does not require the use of another circle theorem. It was a case, which I did not consider during my planning. I did not immediately know how to approach the proof, but I decided to think aloud with my students. After several false starts, I was able to prove the third case. This episode highlights the importance of preparing to notice so that one can notice in-the-moment, and respond better to students' thinking (Mason, 2002). Although the lesson went well for me, it could have gone horribly wrong. Thinking back on this episode, I realised the importance of thinking through the examples or tasks used in my teaching before the lesson so that I can focus more on orchestrating discussions.

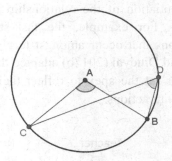

Figure 3. John's case (which I did not consider)

Preparing to notice also has the potential to support teachers to face another challenge—how to orchestrate mathematically productive discussions? This question is important for teachers to consider because it is possible for a well-designed task to be implemented, without realising its design potential. To realise the learning experiences as intended by the tasks, teachers will need to orchestrate the implementation of the tasks in their mathematics classrooms (Tyler, 1949). By recognising that orchestrating discussions requires preparation, they can begin to think about directing their attention to how students (and teacher) discuss mathematics (Yackel & Cobb, 1996). They can consider the patterns of conversations that may hinder or promote student participation (Greeno, 2003; Mehan, 1979; O'Connor & Michaels, 1993) and the type of questions and prompts used in these discourses (Herbal-Eisenmann & Breyfogle, 2005; Mason & Johnston-Wilder, 2006). This is deliberate work, and thus it may be useful to provide a frame for teachers to think about the interactions they intend to orchestrate.

1.3 *What should teachers notice?*

A useful way to think about these classroom interactions is to focus on what the teacher attended to in relation to the interactions between students, content and the task. These interactions can be visualised as a socio-didactical tetrahedron (Rezat & Sträßer, 2012). Referring to Figure 4, we follow Rezat and Sträßer (2012) in seeing each face of the tetrahedron as an instantiation of the relationship between a task and mathematics education. For example, the task-students-teacher face represents the interactions that occur amongst teacher, students and the task. However, Choy and Dindyal (2017b) adapted the tetrahedron model by placing the 'Teacher' at the apex to reflect their focus on how the teacher managed these interactions.

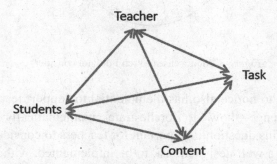

Figure 4. Socio-didactical tetrahedron for using task to orchestrate learning experiences

With the aim of supporting teachers in using tasks to orchestrate learning experiences, a teacher can use the tetrahedron to consider systematically the different facets of task design and implementation.

During task design, a teacher may focus on how the mathematics is encapsulated within the task (teacher-task-content face), and anticipate students' responses to the task (students-task-content face). Similarly, during task implementation, a teacher may listen and observe how students respond to the task before he/she decides on an instructional response (teacher-task-students face). Lastly, the teacher may connect the task to the key mathematical ideas to be taught by a discussion with the students (teacher-content-students face). The tetrahedron provides a way to think about the different interactions, which may feature more prominently at different stages of task design or use in the classroom. To further support teachers to do this work, I now describe a set of task design principles before I highlight how teachers can orchestrate discussions in class.

2 MAD: Mathematics, Activity, and Documentation

Orchestrating learning experiences predicates on the design of a task. As discussed, the design of a mathematical task should provide opportunities for students to engage in mathematical activity. There are many principles for designing a task and they can be broadly encapsulated by the following three principles.

2.1 *The Mathematics Principle*

The first goal of a mathematical task is to engage students in developing mathematical concepts. To do so requires teachers to maintain a clear focus on the mathematics involved (Sullivan et al., 2013). It means, as Anthony and Walshaw (2009) highlight, that teachers can think about how to engage learners in thinking with and about mathematical ideas by identifying the key mathematical concepts, and weaving them into the task (Sullivan et al., 2013). A good starting point is for teachers to understand the common misconceptions, and think about how best to connect learners' current conceptions to the task, so that they can provide opportunities for students to learn from their mistakes (Anthony & Walshaw, 2009).

Building on the Three Point Framework by Yang and Ricks (2013), I suggest that teachers think about the design of a task by considering the concept to be embedded in the task, the errors, misconceptions, or confusion faced by students, and the possible approach to address the confusion (See Figure 5). Together, these three points provide a useful way to think about the design of a task before its implementation.

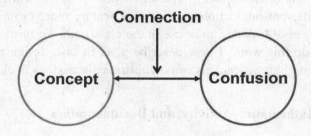

Figure 5. The Mathematics Principle

For example, consider the teaching of fraction-decimal conversion at Primary 4 (Grade 4). The concept targeted in the task may be the idea that common fractions and decimal fractions are different representations of the same number. In terms of students' confusion, it may be their inability to relate fractions with denominators other than 10 to decimals, that is, they may put 1/5 as 0.15 because the digits "1" and "5" appeared in 1/5. A possible approach, which addresses the confusion, is to create tasks where students can relate fractions such as 1/5 to fractions with denominators 10, 100 or 1000.

2.2 *The Activity Principle*

What matters most in a task is the mathematical way of thinking that is embedded in the instructions (Mason & Johnston-Wilder, 2006). The intended mathematical activity of a task, which consists of the actions and thinking, may not correspond to what students do. For example, primary school students may be engaged in drawing different triangles (action), measuring the interior angles in a triangle (action), and calculating the total interior angle sum (action), without realising that their teacher intended them to conjecture about the constant sum of the three angles (action and thinking). As highlighted by Mason and Johnston-Wilder (2006), a mathematical activity may not occur even if the tasks are completed correctly:

> It [mathematical activity] must be more than learners busily getting on with something or producing pages of written work. Although learners may be happily engaged in social interaction through discussion, or fully occupied in using scissors or drawing up tables, they may still not be undertaking mathematical activity. On the other hand, they may be sitting quietly, apparently staring out of the window, and yet be thinking deeply: this could be mathematical activity. For learners' activity to be mathematical, there have to be elements of mathematical thinking... (p. 70)

These elements of mathematical thinking can be seen in tasks, in which learners use their thinking or "natural powers" (Mason & Johnston-Wilder, 2006, p. 74), such as generalising and convincing, to engage with mathematical concepts and relationships. Mathematics tasks should therefore provide students opportunities to consider the mathematical ideas and structures as they act on the tasks; decide on the problem solving approaches to take; and reason about the quality of their own responses (Stein, Grover, & Henningsen, 1996), without always relying on the teacher to provide the directions (Anthony & Walshaw, 2009). As such, students should be the ones engaging in mathematics, and not the teachers.

2.3 *The Documentation Principle*

In order to see whether learners have successfully engaged with learning the mathematical ideas through the mathematical activity initiated by the task, it is essential for learners to make visible their mathematical thinking as much as possible (Lesh, Hoover, Hole, Kelly, & Post, 2000). Therefore, they should have opportunities to communicate their thinking explicitly

(Doerr, 2006) using multiple representations, such as pictures, graphs, numerical data, and algebraic expressions (Anthony & Walshaw, 2009; Sullivan et al., 2013). As a result, teachers can 'see' what and how the students are thinking about the mathematical ideas (Doerr, 2006). The use of representations to express thinking can then provide evidence of student reasoning, which the teacher can choose to take up during the implementation of the task. Therefore, as Anthony and Walshaw (2009) argue, providing students opportunities to communicate using different representations can help them develop their mathematical proficiencies.

3 Orchestrating Discussions

According to Stein, Engle, Smith, and Hughes (2008), using students' correct, partially correct, and incorrect responses to the tasks as the initiator of classroom discourses, and teachers facilitating such interactions in order to shape students' mathematical reasoning is the mark of a well-orchestrated discussion. In order to provide a frame for teachers to think about classroom discussions, Stein et al. (2008) introduce the following five practices:

1. "Anticipating" possible student responses to a task;
2. "Monitoring" their responses when students work with the task;
3. "Selecting" students purposefully to present their work;
4. "Sequencing" their presentations carefully to build up mathematical ideas; and
5. "Connecting" different students' responses to one another, as well as, to relate these responses to the mathematical concepts underlying them. (p. 321)

As Stein et al. (2008) highlight, each of these practices draws from and depends on the practices before it. Smith and Stein (2011) emphasise that these five practices highlight the importance of planning for moderating improvisation during in-the-moment decision making in the classrooms.

The first practice is to anticipate how students may interpret and work on the task mathematically (Smith & Stein, 2011; Stein et al., 2008). It involves interpreting how students may understand the task, how they approach the task, the kind of strategies they use, the possible cognitive obstacles they may face and how these issues relate to the mathematical concepts underlying the task (Smith & Stein, 2011; Stein et al., 2008).

Monitoring student reasoning entails teachers' careful attention to the mathematical aspects of how students work on the task (Schifter, 2001; Stein et al., 2008) when teachers circulate around the classroom. More specifically, it involves teachers focusing on students' mathematical thinking from what they do and say, beyond the surface features of student participation (Stein et al., 2008). For example, instead of noting whether students are actually working on the task, teachers should also listen to students' discussions, in order to assess the mathematical ideas that these students have, and attempt to make sense of students' thinking even when their reasoning is unclear (Schifter, 2001).

Orchestrating a whole-class discussion of the task hinges on a purposeful selection of students' work so that the teacher can get "a particular piece of mathematics on the table" (Lampert, 2001, p. 146), and creates windows of opportunity for students to engage with the concepts (Lampert, 2001; Smith & Stein, 2011). Instead of a laissez-faire approach in selecting students for presentation, a teacher can invite students to volunteer, but purposefully select the students in order to make a mathematical point (Lampert, 2001; Stein et al., 2008). This careful selection ensures that both important mathematical ideas as well as common misconceptions are addressed during the discussion (Stein et al., 2008), and provide opportunities for the teacher to share useful alternative strategies that were not presented by the students (Baxter & Williams, 2009).

Besides selecting students' ideas for presentation, it is also equally important for teachers to sequence the presentations so that they can increase the possibility of attaining the lesson objectives (Smith & Stein, 2011; Stein et al., 2008). For instance, the teacher can start with the most common strategy; or they can begin with a more concrete representation before moving on to a more abstract presentation; or even address a common misconception or difficulty that most students face. They can also consider comparing two contrasting strategies to illustrate the different underlying mathematical perspectives afforded by the task (Smith & Stein, 2011; Stein et al., 2008).

The motivation for carefully selecting and sequencing responses is to lay the groundwork for the teacher to connect these different responses to important mathematical ideas. By directing students' attention to the connections between different strategies, and by shifting their focus from solutions to mathematical ideas, teachers can begin to support students' efforts in understanding the concepts targeted in the lesson (Smith & Stein, 2011; Stein et al., 2008).

As seen from the preceding discussion, these five practices build on one another (see Figure 6), and are critical in supporting teachers to respond appropriately to students' thinking. The five practices assume that there is some mathematical element in students' responses to the tasks designed by teachers, and it is the teacher's role to *listen with* the students in order to make sense of what they have to say. The idea of *listening with students* involves teachers *listening to students*, instead of *listening for some preconceived answers from the students*. When teachers begin to adopt a more listening stance in their teaching, they can begin to attend to, make sense of, and respond better to students' thinking (Choy, Thomas, & Yoon, 2017).

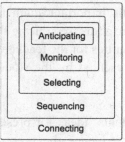

Figure 6. The five practices (Stein et al., 2008, p. 322)

4 Using Typical Problems: Beyond Developing Procedural Skills

Smith and Stein (2011) have argued that tasks, which are of a higher cognitive demand, form the basis for engaging students in doing mathematics. These high-cognitive demand tasks or challenging tasks have rich mathematical connections, and should be used in the mathematics classrooms. However, as Choy and Dindyal (2017b) highlight, while challenging tasks may have their place in the classrooms, they may have too high an entry point for some students, or they may require several lessons for implementation. Moreover, teachers are also mindful about the concurrent need to develop students' procedural fluency as part of their preparation for standardised assessments. Consequently, such tasks may not always feature in the classrooms, and teachers often use them as "rainy day activities". Instead, teachers commonly use examination-type questions together with a teacher-centred teaching approach in Singapore classrooms (Foong, 2009; Ho & Hedberg, 2005). Their preference for using typical problems—standard examination or textbook problems—may reflect teachers' belief that it is "important to prepare students to do well in tests than to implement problem-solving lessons" (Foong, 2009, p. 279), a classroom reality that cannot be ignored.

This raises the question of how teachers can make these mathematical activities more pervasive in the classrooms. I believe that it is possible to tap into the affordances of typical problems and use them to initiate mathematical activities, beyond developing procedural skills. In the following section, I will describe how an experienced mathematics teacher used typical problems to develop both concept and skills, and how the use of typical problems can potentially enhance students' learning experiences.

4.1 *What do typical problems offer?*

Choy and Dindyal (2017b) define typical problems as "standard examination-type questions or textbook-type questions which focus largely on developing procedural fluency and at times, conceptual understanding" (p. 158). As highlighted, these questions require a much shorter time to solve than challenging tasks, and are used frequently in mathematics lessons (Choy & Dindyal, 2017a, 2017b). Given that there are many typical problems in textbooks and other curriculum materials, these questions may hold an untapped potential for orchestrating learning experiences on a day-to-day basis. Therefore, advocating the use of, and the modification of typical problems to create tasks that initiate mathematically meaningful activities (Dindyal, 2018) would then position mathematical learning experiences as an integral part of our mathematics lessons, and not just reserved for the occasional "enrichment" lessons.

4.2 *An illustration: Alice's use of typical problems*

Here, I describe an example of how Alice (pseudonym) used a typical problem to orchestrate a mathematically productive discussion. This vignette provides an example of how typical problems can be used to develop both conceptual understanding and procedural skills. Details of the study can be found in Choy and Dindyal (2017a).

Alice is a Senior Teacher at Coventry Secondary School (pseudonym), a government-funded school with above average performance in the national examinations. As a Senior Teacher, she has a strong mathematical background and has been actively involved in mentoring novice teachers in her school. In this section, I describe Alice's lesson on Matrices for her Secondary Three (Grade 9) students. Alice wanted to provide her students opportunities to apply matrix multiplications to solve contextual problems, and justify if two matrices can be multiplied by checking the order of the matrices. Prior to this lesson, her students had learnt how to multiply two matrices.

For the lesson, Alice modified a typical problem (see Figure 7 for the original question) and used the modified problem as an introductory task to orchestrate a mathematically productive discussion. Alice highlighted that the question could potentially extend students' thinking in terms of expressing the same information in using different representations involving matrices or otherwise. Students often do not have opportunities to express information using different matrices as the matrices are usually given in the question (See Figure 7).

Example 1

[Nov 2013] Teresa and Robert attend the same school. They keep a record of the awards they have earned and the points gained. The matrices show the numbers of awards and the points gained for each award.

$$\begin{array}{c} \\ \text{Teresa} \\ \text{Robert} \end{array} \begin{array}{ccc} \text{Gold} & \text{Silver} & \text{Bronze} \\ \begin{pmatrix} 29 & 10 & 5 \\ 30 & 6 & 8 \end{pmatrix} \end{array} \qquad \begin{array}{c} \\ \text{Gold} \\ \text{Silver} \\ \text{Bronze} \end{array} \begin{array}{c} \text{Points} \\ \begin{pmatrix} 5 \\ 3 \\ 2 \end{pmatrix} \end{array}$$

(a)   Find $\begin{pmatrix} 29 & 10 & 5 \\ 30 & 6 & 8 \end{pmatrix} \begin{pmatrix} 5 \\ 3 \\ 2 \end{pmatrix}$.

(b)   Explain what your answer to (a) represents.

Figure 7. The original typical problem on matrices

As a result, they have limited opportunities to see the connections between arithmetic and matrix multiplication. To achieve her objectives, Alice modified the typical problem as follow:

> Teresa and Robert attend the same school. They keep a record of the awards they have earned and the points gained. Teresa obtained 29 Gold, 10 Silver, and 5 Bronze awards. Robert obtained 30 Gold, 6 Silver, and 8 Bronze awards. They gained 5 points from each Gold award, 3 points for each Silver award, and 2 points for each Bronze award. Find the total number of points that Teresa and Robert gained.

It is interesting to note that Alice did not include any matrix in her modified question. This modification expanded the solution space of the original question. First, students could use an arithmetic approach to solve the question without matrices. Next, students who see the problem as a matrix multiplication problem would first need to formulate the matrices before deciding on the order of matrix multiplication. Hence, Alice could emphasise the connections between matrix multiplication and arithmetic to provide meaning to matrix operations. In so doing, she attempted to develop both conceptual understanding and procedural fluency in the matrix operations. Furthermore, there is more than one way to express the problem in terms of a matrix multiplication. Highlighting the similarities and differences would potentially enriched the students' understanding of the topic.

Alice also demonstrated her recognition of the typical problem's affordances as she orchestrated discussions during the lesson:

1 Alice: (Walks around the class and comes to Student S1.) Can you write this for me on the board?

2 S1:   Ok. (Walks to the whiteboard and writes the following:
$$T = 5 \times 29 + 3 \times 10 + 2 \times 5 = 185$$
$$R = 5 \times 30 + 3 \times 6 + 2 \times 8 = 184)$$

3 Alice: (Walks around while waiting for Student S1 to finish writing.) Ok. Most of you have written what [Student S1] has written. 5 points for 29 gold, 3 points for 10 silver and 2 points for 5 bronze. Most of you have written in this manner. The last few days, we have been talking about matrices, right? Would you like to convert this to a matrix problem?
(Looks at Student S2) Have you written it in matrix form? (Student S2 nods and Alice goes over to look at his answers.) Okay. Can you write your answer on the board?

4 S2:   (Walks to the board and writes the following.)
$$T = (29 \quad 10 \quad 5) \times \begin{pmatrix} 5 \\ 3 \\ 2 \end{pmatrix} = 185 \text{ and}$$
$$R = (30 \quad 6 \quad 8) \times \begin{pmatrix} 5 \\ 3 \\ 2 \end{pmatrix} = 184$$

5 Alice: Any other answers from [Student S2's] answer? (Walks around the class and selects Student S3's answer) Can you write this on the board?

6 S3:   (Walks to the board and writes the following.)

$$\begin{pmatrix} 29 & 10 & 5 \\ 30 & 6 & 8 \end{pmatrix} \begin{pmatrix} 5 \\ 3 \\ 2 \end{pmatrix} = \begin{pmatrix} 29 \times 5 + 10 \times 3 + 5 \times 2 \\ 30 \times 5 + 6 \times 3 + 8 \times 2 \end{pmatrix}$$

$$= \begin{pmatrix} 185 \\ 184 \end{pmatrix}$$

7   Alice: Thank you all three of you. [Student S1] has written using an arithmetic method. Most of you have written in this manner. This one comes very naturally to you, ok? [Student S2] has written Robert and Theresa's award separately. He has tried to use the matrix method, (points to Student S1's solution.) Something like this, ok? Let's check whether the order of matrix is correct or not.

(Alice goes through the method of matrix multiplication and gets the class to check the order of Student S2's matrices.)

… Ok. Student S3 has written Robert's and Theresa's together so that you only write this matrix once (points to the column matrix [5 3 2]). Don't need to write two times, correct or not? See. Over here. You have to write two times but here, [Student S3] only has to write it once. Let's check the order again…

8   Alice: (After a short time) I would like to bring this problem a little bit further. Notice that Student S3 presented the information this way. Is there another way to represent the same information?

(After some time, Student S4 highlights another possible way.)

Here, Alice orchestrated a mathematically productive discussion (Smith & Stein, 2011). By beginning with an arithmetic solution, Alice connected Student S1's arithmetic operations to matrix multiplications through the sequencing of Student S2's and Student S3's matrix solutions. The reason for using a single matrix multiplication (Student S3's solution) was also made explicit when Alice moved from Student S2's solution to Student S3's using a matrix approach (Line 7) before she highlighted the different ways to express the given information as matrices of different orders (Line 8), which was an important idea for the lesson.

Alice's case illustrate the possibility of using typical problems to teach for conceptual understanding in at least two ways. First, she noticed the potential of the task to initiate students' thinking about matrices, by focusing their attention on *why* it makes sense to define matrix multiplication in the way mathematicians did. Next, she demonstrated how Smith's and Stein's (2011) five practices can be manifested using a typical problem. In the actual lesson, Alice used a sequence of typical problems, such as the one described here, to orchestrate a discussion using the same moves—monitoring, selecting, sequencing, and connecting—in rapid succession. Her use of these four practices was made feasible by the use of typical problems, which generally takes a shorter time to complete. What makes Alice's case interesting is how she has used typical problems to develop students' procedural fluency, while maintaining a focus on engaging students to think more deeply about matrices. In other words, the task (the typical problem) has been used to initiate a more meaningful mathematical activity, beyond its usual use to hone skills.

5 From Task to Activity: Some Concluding Remarks

In this chapter, we see that it is important to distinguish a task from its activity because of how students may engage with a given task. For students to experience the task and engage in the mathematical activities afforded by the tasks, it is crucial for teachers to think about the design and implementation of tasks in the classrooms. One way to do this is to notice the affordances of a task, and use the task as a means to address students' confusion when learning a concept. In this way, teachers begin to think about the design of a task, beyond its surface features (such as novelty, graphics, etc.). The design is just the first step towards enhancing students' learning experiences. More importantly, how teachers orchestrate mathematically productive discussions can significantly change the mathematical activities engaged by students. The same task, for example the typical problem, can be implemented as a drill and practice question, or as Alice has demonstrated, a means to engage students in thinking about matrices. Finally, it is not about designing a mathematically rich task set in a real world context or a typical problem that matters, but how students engage with the mathematics, which makes a difference.

Acknowledgement

Part of this chapter refers to data from the research project "Portraits of teacher noticing during orchestration of learning experiences in the mathematics classrooms" (OER 03/16 CBH), funded by the Office of Educational Research (OER), National Institute of Education (NIE), Nanyang Technological University, Singapore, as part of the NIE Education Research Funding Programme (ERFP). The views expressed in this chapter are the author's and do not necessarily represent the views of NIE.

I would like to thank Associate Professor Dindyal Jaguthsing, my co-Principal Investigator (OER 03/16 CBH), for his insights into task design during our project discussions.

References

Anthony, G., & Walshaw, M. (2009). *Effective pedagogy in mathematics*. Geneva: International Bureau of Education.

Ball, D. L., & Forzani, F. M. (2009). The work of teaching and the challenge for teacher education. *Journal of Teacher Education, 60*(5), 497-511. doi:10.1177/0022487109348479

Baxter, J. A., & Williams, S. (2009). Social and analytic scaffolding in middle school mathematics: Managing the dilemma of telling. *Journal of Mathematics Teacher Education, 13*(1), 7-26. doi:10.1007/s10857-009-9121-4

Choy, B. H. (2014). Teachers' productive mathematical noticing during lesson preparation. In C. Nicol, P. Liljedahl, S. Oesterle, & D. Allan (Eds.), *Proceedings of the Joint Meeting of PME 38 and PME-NA 36* (Vol. 2, pp. 297-304). Vancouver, Canada: PME.

Choy, B. H. (2016). Snapshots of mathematics teacher noticing during task design. *Mathematics Education Research Journal, 28*(3), 421-440. doi:10.1007/s13394-016-0173-3

Choy, B. H., & Dindyal, J. (2017a). Noticing affordances of a typical problem. In B. Kaur, W. K. Ho, T. L. Toh, & B. H. Choy (Eds.), *Proceedings of the 41st Conference of the International Group for the Psychology of Mathematics Education* (Vol. 2, pp. 249-256). Singapore: PME.

Choy, B. H., & Dindyal, J. (2017b). Snapshots of productive noticing: orchestrating learning experiences using typical problems. In A. Downton, S. Livy, & J. Hall (Eds.), *40 years on: we are still learning! Proceedings of the 40th annual conference of the Mathematics Education Research Group of Australasia* (pp. 157-164). Melbourne: MERGA.

Choy, B. H., Thomas, M. O. J., & Yoon, C. (2017). The FOCUS framework: characterising productive noticing during lesson planning, delivery and review. In E. O. Schack, M. H. Fisher, & J. A. Wilhelm (Eds.), *Teacher noticing: bridging and broadening perspectives, contexts, and frameworks* (pp. 445-466). Cham, Switzerland: Springer.

Dindyal, J. (2018). Affordances of Typical Problems. In P.C. Toh, & B. L. Chua (Eds.), *Mathematics Instruction: Goals, Tasks and Activities* (pp. 33-48). Singapore: World Scientific

Doerr, H. M. (2006). Examining the tasks of teaching when using students' mathematical thinking. *Educational Studies in Mathematics, 62*(1), 3-24. doi:10.1007/s10649-006-4437-9

Foong, P. Y. (2009). Review of research on mathematical problem solving in Singapore. In K. Y. Wong, P. Y. Lee, B. Kaur, P. Y. Foong, & S. F. Ng (Eds.), *Mathematics education: the Singapore journey* (pp. 263 - 300). Singapore: World Scientific.

Greeno, J. G. (2003). Situative research relevant to standards for school mathematics. In J. Kilpatrick, W. G. Martin, & D. Schifter (Eds.), *A research companion to principles and standards for school mathematics* (pp. 304-332). Reston, VA: National Council of Teachers of Mathematics.

Herbal-Eisenmann, B. A., & Breyfogle, M. L. (2005). Questioning Our "Patterns" of Questioning. *Mathematics teaching in the middle school, 10*(9), 484-489.

Ho, K. F., & Hedberg, J. G. (2005). Teachers' pedagogies and their impact on students' mathematical problem solving. *The Journal of Mathematical Behavior, 24*(3-4), 238-252. doi:10.1016/j.jmathb.2005.09.006

Lampert, M. (2001). *Teaching problems and the problems of teaching*. New Haven & London: Yale University Press.

Lesh, R., Hoover, M., Hole, B., Kelly, A., & Post, T. (2000). Principles for developing thought-revealing activities for students and teachers. In A. Kelly & R. Lesh (Eds.), *Handbook of research design in mathematics and science education* (pp. 591-646). Mahwah, New Jersey: Lawrence Erlbaum Associates.

Mason, J. (2002). *Researching your own practice: The discipline of noticing*. London: RoutledgeFalmer.

Mason, J. (2010). Attention and intention in learning about teaching through teaching. In R. Leikin & R. Zazkis (Eds.), *Learning through teaching mathematics: Development of teachers' knowledge and expertise in practice* (pp. 23-47). New York: Springer.

Mason, J., & Johnston-Wilder, S. (2006). *Designing and using mathematical tasks*. United Kingdom: Tarquin Publications.

Mehan, H. (1979). *Learning lessons: Social organization in the classroom.* Cambridge, MA: Harvard University Press.

Ministry of Education. (2013). *Primary mathematics teaching and learning syllabus.* Singapore: Curriculum Planning and Development Division.

O'Connor, M. C., & Michaels, S. (1993). Aligning academic task and participation status through revoicing: Analysis of a classroom discourse strategy. *Anthropology & Education Quarterly, 24*(4), 318-335.

Rezat, S., & Sträßer, R. (2012). From the didactical triangle to the socio-didactical tetrahedron: artifacts as fundamental constituents of the didactical situation. *ZDM Mathematics Education, 44*(5), 641-651. doi:10.1007/s11858-012-0448-4

Schifter, D. (2001). Learning to see the invisible: What skills and knowledge are needed to engage with students' mathematical ideas? In T. Wood, B. S. Nelson, & J. Warfield (Eds.), *Beyond classical pedagogy: Teaching elementary school mathematics* (pp. 109-134). Mahwah, NJ: Lawrence Erlbaum Associate, Inc.

Sherin, M. G., Jacobs, V. R., & Philipp, R. A. (Eds.). (2011). *Mathematics teacher noticing: Seeing through teachers' eyes.* New York: Routledge.

Smith, M. S., & Stein, M. K. (2011). *5 practices for orchestrating productive mathematics discussions.* Reston, VA: National Council of Teachers of Mathematics Inc.

Stein, M. K., Engle, R. A., Smith, M. S., & Hughes, E. K. (2008). Orchestrating productive mathematical discussions: Five practices for helping teachers move beyond show and tell. *Mathematical Thinking and Learning, 10*(4), 313-340. doi:10.1080/10986060802229675

Stein, M. K., Grover, B. W., & Henningsen, M. (1996). Building student capacity for mathematical thinking and reasoning: An analysis of mathematical tasks used in reform classrooms. *American Educational Research Journal, 33*(2), 455-488.

Sullivan, P., Clarke, D., & Clarke, B. (2013). *Teaching with tasks for effective mathematics learning.* New York: Springer.

Tyler, R. W. (1949). *Basic Principles of Curriculum and Instruction.* Chicago: The University of Chicago Press.

Yackel, E., & Cobb, P. (1996). Sociomathematical norms, argumentation, and autonomy in mathematics. *Journal for Research in Mathematics Education, 27*(4), 458-477.

Yang, Y., & Ricks, T. E. (2013). Chinese lesson study: Developing classroom instruction through collaborations in school-based teaching research group activities. In Y. Li & R. Huang (Eds.), *How Chinese teach mathematics and improve teaching* (pp. 51-65). New York: Routledge.

Chapter 3

Affordances of Typical Problems

Jaguthsing DINDYAL

Teachers use mathematical tasks to develop students' understanding of mathematics. Typical problems are mathematical tasks that are readily available to teachers and are often used quite procedurally for developing students' skills. What are the affordances of typical problems for developing conceptual fluency? In this chapter, I argue that teachers can perceive the affordances of using typical problems to develop conceptual fluency when they are able to modify typical problems in various ways. I suggest several strategies for modifying typical problems which can be used productively in mathematics lessons.

1 Introduction

Mathematical tasks are central to students' learning because "tasks convey messages about what mathematics is and what doing mathematics entails" (NCTM, 1991, p. 24). Accordingly, teachers use a multitude of mathematical tasks in their day-to-day practice to help students to grasp the mathematical content. Over the years, several terms have been used to describe mathematical tasks. Amongst others, we have: worthwhile mathematical tasks (NTCM, 1991), challenging tasks (Sullivan et al., 2014), high-level tasks (Henningsen & Stein, 1997), open-ended tasks (Zaslavsky, 1995), and rich mathematical tasks (Grootenboer, 2009). While acknowledging the benefits of using such tasks, we must also be vigilant of the fact that the task with which "students actually engage may or may not be the same task that the teacher announced at the outset" (Stein, Grover, & Henningsen, 1996, p. 462). Furthermore, teachers face several issues in selecting, adapting and actually implementing such tasks. In this chapter, I propose typical problems as an alternative to these types of tasks. Typical problems from past examination papers, textbooks, etc. provide a rich resource for teachers. To perceive the affordances of typical problems, teachers should be able to modify typical problems for classroom use. Accordingly, this chapter focuses on how teachers can select, modify and use typical problems in the teaching of mathematics.

1.1 *Mathematical tasks*

Watson and Thompson (2015) refer to a task as the written presentation of a planned mathematical experience for a learner, which could be one action or a sequence of actions that form an overall experience. Thus, a task could consist of anything from a single problem, or a textbook exercise, to a complex interdisciplinary exploration. Consider this mathematical task:

Task A
Unit cubes are used to make larger cubes of other sizes. The surface area of each of the large cubes are painted and then disassembled into the original unit cubes. For each large cube, investigate how many of the unit cubes are painted on three faces, two faces, one face, and no faces? Describe the patterns you observe.

Considering $2 \times 2 \times 2$, $3 \times 3 \times 3$, $4 \times 4 \times 4$ and $5 \times 5 \times 5$ cubes, and generalising to the $n \times n \times n$ cube, the following patterns can be observed:

Number of unit cubes with paint on 3 faces: 8, 8, 8, 8 ..., 8
Number of unit cubes with paint on 2 faces: 0, 12, 24, 36 ... $12(n-2)$
Number of unit cubes with paint on 1 face: 0, 6, 24, 54 ... $6(n-2)^2$
Number of unit cubes with paint on no faces: 0, 1, 8, 27 ... $(n-2)^3$

We note that the patterns generate constant, linear, quadratic and cubic functions respectively. In the absence of any manipulative, the task involves the students visualizing the larger cubes and sequentially recording their observation to find meaningful patterns. The teacher can let the students struggle on their own to get the answers. Additionally, the teacher may ask more challenging questions such as: can the number of cubes with paint on one face be 1800 for any large cube? How many unit cubes will you need to make a large cube that will have 729 cubes with no paint on any face? This mathematical task can be considered as a *rich task* as it provides many avenues for exploration and extension. The task may take up a whole lesson for proper implementation. On the other hand, many students may not come to a generalization stage even if they can notice a pattern. Some students may not understand the problem at all and be at a loss. The teacher can explore this task in various ways with the students by providing some guiding questions: For a $3 \times 3 \times 3$ cube, which unit cubes will be painted on three faces? Which unit cubes will be painted on only two faces? Which unit cubes will be painted on only one face? Which unit cubes will not be painted at all? It is to be noted that not every student will benefit from these scaffolding questions. There are several other issues, for example, the teacher may simply focus on trivial matters like getting to the final answer and implement this task in a very routine way.

Task B
A bag contains 5 identical red balls and 3 identical black balls. Two balls are taken from the bag at random without replacement.
i) Draw a tree diagram to show the probabilities of the possible outcomes.
ii) Find as a fraction in its simplest form, the probability that
  a) the first ball taken is red,
  b) both balls are black,
  c) one ball is red and the other is black.

This particular *typical problem* is modelled on a similar problem from the O-level specimen paper. The first part of this task can be solved easily by drawing a tree diagram, which is a skill that students practise quite a lot at the secondary level. The second part has three sub-parts, each of which can be solved using the tree diagram from the first part or independently by considering all outcomes. The focus here is not so much on solving the task but much more on the fact that this task is what can be termed as a typical problem, as it has a fairly straightforward answer with a very moderate difficulty level, that students have practised previously.

2 Typical Problems

Typical problems are the simple problems that teachers find in commonly available resources such as textbooks and past examination papers. Choy and Dindyal (2017b) consider typical problems as:

> … standard examination-type questions or textbook-type questions which focus largely on developing procedural fluency and at times, conceptual understanding... These questions can be solved by students in a much shorter time than challenging tasks, and are used frequently in mathematics lessons. Given the omnipresence of such questions in textbooks and other curriculum materials, we see typical problems as an untapped resource, which can be used to orchestrate learning experiences on a day-to-day basis. Using tasks developed from typical problems to orchestrate learning experiences would position mathematical learning experiences as an integral part of our mathematics lessons, and not just reserved for the occasional "enrichment" lessons. (p. 158)

A typical problem such as Task B above can be solved procedurally. As such, the teacher can direct students' attention on the skills for solving the problem but can this problem be used otherwise? In particular, can it be used for developing conceptual fluency? While it is expected that typical problems can be used for developing procedural skills, it is not always expected that such problems could be used for developing conceptual fluency. Choy and Dindyal (2017a, 2017b) have demonstrated how a teacher Alice used typical problems to develop both procedural skills as well as conceptual fluency. In other words, typical problems can provide *affordances* for the typical problem to be used to developing conceptual fluency.

3 Affordances

The term *affordances* was coined by the perceptual psychologist Gibson (1986). Taking a broader perspective, Gibson claimed that, "The affordances of the environment are what it offers the animal, what it provides or furnishes, either for good or ill." (p.127) His basic thesis is that an affordance of an object exist in relation to an observer; the affordance of that object does not change as the need of the observer changes; the observer may or may not perceive or attend to the affordance according to his needs, but the affordance being invariant is always there to be perceived. Thus, he added that, "An affordance is not bestowed upon an object by a need of an observer and his act of perceiving it" (Gibson, 1986, pp. 138-139). If we apply this idea to typical problems and consider the teacher to be the observer then we can state that: (1) an affordance for using a typical problem exists relative to the action and capabilities of the teacher; (2) the existence of the affordance is independent of the teacher's ability to perceive it; (3) the affordance does not change as the needs and goals of the teacher change. Gibson also highlighted that affordances in relation to an observer could be positive or negative which in our context may lead to productive or less productive use of the problems in class by the teacher. Hence, to perceive the affordances of a typical problem means to be able to *notice* the characteristics of the task in relation to the particular understandings of the related concept in order to adapt the task for use in classrooms (see Choy & Dindyal, 2017a, 2017b). But how can we help a teacher to notice the affordances of a typical problem?

If we assume that the teacher knows the mathematical content to be taught then here are a few ideas that can be considered:

1. the teacher should be able to identify what exactly the student has to learn in terms of the concepts, conventions, results, techniques, and processes (see Backhouse, Haggarty, Pirie, & Stratton, 1992);

2. the teacher should be able to unpack the curriculum, looking at the content as *nested components*, each made up of tasks or a sequence of tasks (typical problems) within lessons and lessons within a sequence of content units, which are ultimately part of the mathematics curriculum;

3. the teacher should be able to select appropriate tasks (typical problems) which he or she can design, or use existing ones and modify the typical problems to suit their teaching needs;

4. the teacher should be able to review these tasks or typical problems after having used these in their class.

4 How to Modify Typical Problems?

To notice the affordances of a typical problem, a teacher should be able to modify the typical problem in various ways. The modification of typical problems can surface approaches to implementation of these problems in class either for developing procedural skills or for developing conceptual fluency. In this chapter, I will focus on the approaches to modify a typical problem using Task B as an example to illustrate a few cases. There are several ways to modify a typical problem such as Task B.

1. Changing the numbers in the typical problem

Modified Problem 1	Principle of modification
A bag contains 20 identical red balls and 10 identical black balls. Two balls are taken from the bag at random without replacement.	The problem is exactly the same as the original problem except for the change of the number of balls.

i) Draw a tree diagram to show the probabilities of the possible outcomes.
ii) Find as a fraction in its simplest form, the probability that
   a) the first ball taken is red,
   b) both balls are black,
   c) one ball is red and the other is black.	

One common way to modify a typical problem involves changing the numbers in the problem. This change should not be an arbitrary change of numbers but rather numbers that will make sense in the context of the problem and the intended use of the problem in the class. Some students find it harder to work out problems involving larger numbers.

2. Changing the context of the problem

Modified Problem 2	Principle of modification
Five red cards and three black cards are selected from a deck of playing cards. The cards are shuffled and placed face down on the table. Two cards are then chosen without replacement.	The only change is in the context which is now involving cards instead of balls.

i) Draw a tree diagram to show the probabilities of the possible outcomes.
ii) Find as a fraction in its simplest form, the probability that
  a) the first card taken is red,
  b) both cards are black,
  c) one card is red and the other is black.	

Another way to modify a typical problem involves changing the context of the problem. The context can be made more accessible or simpler for students but can also be made more complex and demanding if the need arises for highlighting particular ideas in class. The change of the context from one involving balls in Task B to one involving cards in Modified Problem 2 does not necessarily make the problem harder or easier. This may be a useful strategy if the original typical problem comes from a different socio-cultural context.

3. Changing specific parts of the problem

Modified Problem 3	Principle of Modification
A bag contains five identical red balls and three identical black balls. One ball is chosen at random. Find as a fraction in its simplest form, the probability that this first ball is red.	Changing what is asked to be done by reducing the number of parts.

One way to modify a typical problem, involves changing what is asked to be done by reducing the number of parts and making the problem simpler as shown in Modified Problem 3 above. This type of modification of the typical problem can also be carried out in a way that still retain an element of challenge for the students, as shown in Modified Problem 4.

Modified Problem 4	Principle of Modification
A bag contains five identical red balls and three identical black balls. Two balls are taken at random from the bag without replacement. Find as a fraction in its simplest form if one ball is red and the other is black.	Changing what is asked to be done by reducing the number of parts.

Modified Problem 4 is not as direct as Modified Problem 3, although both of these problems involved reducing the number of parts in the problem. Yet another way to modify a typical problem involves adding a part (usually harder) to the existing number of parts in the problem as shown in Modified Problem 5 below.

Modified Problem 5	Principle of Modification
A bag contains five identical red balls and three identical black balls. One ball is chosen, it is not replaced, then another ball is chosen.	Changing what is asked by adding an extra part.

i) Draw a tree diagram to show the probabilities of the possible outcomes.
ii) Find as a fraction in its simplest form, the probability that
    a) the first ball taken is red,
    b) both balls are black,
    c) one ball is red and the other is black,
    d) one ball is red and the other is green.

4. Changing one or more conditions in the problem

Modified Problem 6	Principle of Modification

A bag contains five identical red balls and three identical black balls. One ball is chosen, it is replaced, then another ball is chosen.
 	
i) Draw a tree diagram to show the probabilities of the possible outcomes.
ii) Find as a fraction in its simplest form, the probability that
   a) the first ball taken is red,
   b) both balls are black,
   c) one ball is red and the other is black.
	Changing one condition: "without replacement" to "is replaced".

A typical problem can also be modified by changing one or more conditions in the problem. For example, the condition "without replacement" from Task B, is changed to "is replaced" in Modified Problem 6. This type of modification is quite subtle and students should be exposed to such modifications for them to understand how the solution process is affected just by the changes in specific conditions in the problem.

5. Connecting with another topic in mathematics

Modified Problem 7	Principle of Modification
A bag contains n identical red balls and $(n-2)$ identical black balls. One ball is chosen, it is replaced, then another ball is chosen. As a fraction in its simplest form, the probability that both balls are black is $\frac{1}{9}$. Form an equation in n and show that it reduces to $5n^2 - 28n + 32 = 0$. Solve this equation to find n.	Combining with another topic in mathematics, namely "solution of quadratic equations".

A typical problem can be modified by making connections with other topics in mathematics. For example, Modified Problem 7 above includes the same ideas from Task B but involves the solving of a quadratic equation.

6. Creating open-ended problems

Modified Problem 8	Principle of Modification
A bag contains identical balls, some of which are red and some are black. The probability of choosing a black ball from the bag is $\frac{2}{5}$. What could be the number of red balls in the bag?	Changing what is given and what is to be found. Creating an open-ended problem.

A typical problem can be modified by changing what is given and what is to be found and/or creating an open-ended problem. In Modified Problem 8, a student will have to work backwards from the given probability to find the number of red balls. This problem is also open-ended as there may be more than one possible correct solution. The same ideas are illustrated in Modified Problem 9 as well.

Modified Problem 9	Principle of Modification
A bag contains identical balls, some of which are red and some are black. One ball is chosen, then another ball is chosen after replacing the first ball. The probability of choosing two black balls from the bag is $\frac{4}{25}$. What is the number of red balls in the bag?	Creating an open-ended problem.

7. Providing opportunities to generalize

Modified Problem 10	Principle of Modification
A bag contains identical balls, of which n are red and $(n\text{-}2)$ are black. One ball is chosen, it is not replaced then another ball is chosen. Find, in terms of n, the probability that a) the first ball taken is red, b) both balls are black, c) one ball is red and the other is black.	Creating an opportunity to generalize.

Task B can be modified to provide students with opportunities to generalize as illustrated in Modified Problem 10. Not all problems can provide students with opportunities to generalize. However, students should have the opportunity to practise generalization whenever it is possible to do so.

8. A combination of the previous methods of modification

Modified Problem 11	Principle of Modification
Ten red cards and five black cards are selected from a deck of playing cards. The cards are shuffled and placed face down on the table. One card is chosen, it is not replaced, then another card is chosen.	A combination of the above methods of modification: change numbers, change context, change condition, etc.
i) Draw a tree diagram to show the probabilities of the possible outcomes. ii) Find as a fraction in its simplest form, the probability that a) the first card taken is red, b) both cards are black, c) one card is red and the other is black.	

The previous methods for modifying typical problems can be combined in various ways to create new problems. For example, Modified Problem 11 demonstrates a combination of methods for modifying a typical problem.

5 Concluding Remarks

To summarise, the above strategies for modifying typical problems are not exhaustive. There may be other ways in which problems can be modified (see Kilpatrick, 1987; Abedi & Lord, 2001). However, the strategies listed above, do provide a good range of possible ways, either individually or in combination with other strategies, to modify typical problems. It is important to note that problem modification is not something entirely new. Teachers do it all the time but typical problems provide more flexibility as compared to the so-called rich tasks. There are several reasons for using typical problems:

1. Typical problems are more accessible to teachers, as compared to rich tasks. Sourcing for relevant rich tasks or designing and modifying original rich tasks can be a challenge for the average secondary school teacher.
2. Rich tasks are usually quite rigid and provide less flexibility for modification as compared to typical problems.
3. Rich tasks usually connect several ideas from different content domains, which is a good thing. However, such tasks are quite challenging for the average student. On the other hand, typical problems can be designed or modified to focus on specific concepts for students to gain mastery before embarking on rich tasks.
4. Rich tasks take much longer time for implementation as compared to typical problems and provide a challenge to teachers who aim to complete the syllabus requirements and prepare students for the examinations. A sequence of well-designed (or modified) typical problems can be more focused on the specific concepts that are taught and can be as such used for developing both procedural skills and conceptual fluency.

To modify typical problems for use in mathematics classrooms, as detailed above, I suggest the following procedure. First, focus on the given typical problem; second, focus on the kind of modification that suits your implementation needs; third, focus on the implementation of the problem in your class, and finally after the implementation, take time to review the problem and its use in the classroom and whether to keep it or to further modify it for subsequent use.

5.1 *Focusing on the given problem*

1. Check the lesson unit on which this problem is based and the expected Learning Outcomes.
2. Solve the problem (in multiple ways, if possible).

3. Check the problem for the concepts on which it is based. Do your students have the necessary resources to solve the problem? Is the problem within the requirements of the syllabus?

4. Check the problem for the skills and sub-skills that it is supposed to test.

5. Check the problem for any results, techniques or conventions that are used by students in the solution of the problem.

6. Check the problem for the kinds of processes that are emphasized in the solution.

7. Check the problem for the kind of structure that is already present in the statement of the problem: diagrams, charts, graphs, tables, parts and subparts, etc.

8. Check the problem for the language: action verbs, key words and phrases, technical/mathematical terms, connectives, etc.

9. Look out for the kind of affordances this problem provide for enhancing your students' learning experience.

5.2 *Focusing on the modification of the problem*

1. What kinds of skills do you want to elicit from the students? Is it a recall of some definition? Do they have to draw something? Do they have to calculate, solve or prove?

2. Think of whether you wish to make the problem harder or easier. Think of reducing or increasing the structure in the problem by drawing diagrams, adding parts or subparts, using a table, or using a graph or a chart.

3. Select an appropriate context.

4. Use the appropriate action or direction verb: Find, Calculate, Evaluate, List, Draw, Sketch, Show, Prove, Obtain, Express, Deduce, Shade, State, Solve, Estimate, Tabulate, Explain how, Explain why, Plot, Write down, Factorise, Simplify, etc.

5. Use appropriate keywords: Not, All, And, Or, Some, If...then, Because, At least, At the most, If and only if, Sometimes, Always, Never, Equal to, Less than or equal to, The least value, etc.

6. Think of changing the numbers, the context, what is given and what is to be found.
7. Think of creating an open-ended problem or integrating with another branch of mathematics.
8. If you wish your students to notice something, think of keeping something fixed and changing or modifying something.
9. Think of how the procedural skills and conceptual knowledge can be elicited in the orchestration of the lesson.
10. Think of how this problem either as a stand-alone problem or in combination with other problems provides an excellent way to test the students' procedural skills and conceptual knowledge.

5.3 *Focusing on the implementation of the problem*

1. Think of when you will use the problem in class. Focus on the orchestration of the lesson.
2. Anticipate the kinds of solutions you expect from your students. Which solutions will you highlight for the rest of the class?
3. Think of the kinds of modifications of the problem that you will use in class.
4. Think of the kinds of questions that you will ask the students to deepen their understanding.
5. Think of other typical problems (and in which sequence) that you can use alongside this problem to enhance the students' learning.

5.4 *Reflecting on the use and the modification of the problem*

1. It is good to carefully review the use of the typical problem in the lesson, either on its own or in conjunction with other problems.
2. Was this problem used optimally to enhance your students' learning? What other affordances do the problem or the modification provide that can be explored further?

The affordances of typical problems for classroom use by mathematics teachers can only be perceived if seen in the light of possible avenues for modifications. Of course, teachers can only use the typical problem in a procedural way and be contented with its use. However, the possible ways for modification of typical problems can provide teachers with alternative ways to think about the implementation of the problem in the classroom. Typical problems can be modified to develop conceptual fluency as well, either individually or when used in an appropriate sequence in the classroom.

References

Abedi, J., & Lord, C. (2001). The language factor in mathematics tests. *Applied Measurement in Education, 14*(3), 219-234.

Backhouse, J., Haggarty, L., Pirie, S., & Stratton, J. (1992). *Improving the learning of mathematics*. London, UK: Cassell.

Choy, B. H., & Dindyal, J. (2017a). Noticing affordances of a typical problem. In B. Kaur, W. K. Ho, T. L. Toh, & B. H. Choy (Eds.), *Proceedings of the 41st Conference of the International Group for the Psychology of Mathematics Education* (Vol. 2, pp. 249-256). Singapore: PME.

Choy, B. H., & Dindyal, J. (2017b). Snapshots of productive noticing: orchestrating learning experiences using typical problems. In A. Downton, S. Livy, & J. Hall (Eds.), *40 years on: we are still learning!* Proceedings of the 40th annual conference of the Mathematics Education Research Group of Australasia (pp. 157-164). Melbourne: MERGA.

Gibson, J. J. (1986). *The ecological approach to visual perception*. Hillsdale, New Jersey: Lawrence Erlbaum Associates, Publisher.

Grootenboer, P. (2009). Rich mathematical tasks in the Maths in the Kimberly (MITK) project. In R. Hunter, B. Bicknell, & T. Burgess (Eds.), *Crossing divides: Proceedings of the 32nd annual conference of the Mathematics Education Research Group of Australasia* (Vol. 1, pp. 696-699). Palmerston North, NZ: MERGA.

Henningsen, M., & Stein, M. K. (1997). Mathematics tasks and student cognition: Classroom-based factors that support and inhibit high-level mathematical thinking and reasoning. *Journal for Research in Mathematics Education, 28*(5), 524-549.

Kilpatrick, J. (1987). Problem formulating: Where do good problems come from? In A. H. Schoenfeld (Ed.), *Cognitive science and mathematics education* (pp. 123-147). Hillsdale, NJ: Lawrence Erlbaum Associates.

National Council of Teachers of Mathematics (NCTM). (1991). *Professional Standards for Teaching Mathematics*. Reston, VA: NCTM.

Stein, M. K., Grover, B. W., & Henningsen, M. (1996). Building student capacity for mathematical thinking and reasoning: An analysis of mathematical tasks used in reform classrooms. *American Educational Research Journal, 33*(2), 455-488.

Sullivan, P., Askew, M., Cheeseman, J., Clarke, D., Mornane, A., Roche, A., & Walker, N. (2014). Supporting teachers in structuring mathematics lessons involving challenging tasks. *Journal of Mathematics Teacher Education, 18*(2), 123-140. doi:10.1007/s10857-014-9279-2

Watson, A., & Thompson, D. (2015). Design issues related to text-based tasks. In A. Watson, & M. Ohtani (Eds.), *Task design in mathematics education* (pp. 143-190). NY: Springer.

Zaslavsky, O. (1995). Open-ended tasks as a trigger for mathematics teachers' professional development. *For the Learning of Mathematics, 15*(3), 15-20.

Chapter 4

Mathematical Tasks Enacted by Two Competent Teachers to Facilitate the Learning of Vectors by Grade Ten Students

Berinderjeet KAUR   Lai Fong WONG   Chong Kiat CHEW

Teachers use mathematical tasks to engage students in activities and facilitate learning. Individual teachers arrange instruction very differently, depending on what they are teaching, and students respond to instruction very differently, depending on the structure and demands shaped by tasks enacted in the classroom. In this chapter, we examine mathematical tasks enacted by two competent teachers to facilitate the learning of vectors by grade 10 students. There were similarities and differences in the tasks they used. To match the goals of their lessons, they used similar types of tasks. However, to match the interest and ability of their students their tasks differed in context and cognitive demands. The source of the tasks also differed—one teacher used a collection of tasks that were developed over a period of time; the other used and adapted tasks from the textbook in addition to tasks that were created to supplement the achievement of lesson goals.

1 Introduction

The basic aim of a mathematics lesson is for learners to learn some aspect of mathematics, be it a concept, skill or mathematical process. To do this, teachers engage students in activities using tasks (Christiansen & Walther, 1986). Individual teachers arrange instruction very differently, depending on what they are teaching, and students respond to instruction very differently, depending on the structure and demands shaped by tasks enacted in the classroom (Shimizu, Kaur, Huang, & Clarke, 2010). The tasks that teachers assign can determine how students come to understand what is taught. Doyle argues that tasks influence learners by directing their attention to particular aspects of content and by specifying ways of processing information (Doyle, 1983, p.161). Therefore, to achieve quality mathematics instruction the role of mathematical tasks to stimulate students' cognitive processes is crucial (Hiebert & Wearne, 1993).

A lesson often comprises several phases, like warming up, building skills, developing conceptual understanding, practice leading to procedural fluency and making connections within the topic. For each of these phases, teachers plan activities comprising suitable tasks that achieve specific lesson goals. When a sequence of lessons for a topic of instruction is documented, the data allows one to examine in its entirety how the teacher sequenced the tasks and what were the purposes of the tasks.

When introducing a new topic, the teacher may want to review prior knowledge, if appropriate, or make connections to knowledge in other topics or topics in other curriculum subjects. When beginning a lesson, similarly the teacher may use warm-up tasks or review tasks to refresh learners' memories of work done in the past lesson (Mason & Johnston-Wilder, 2006). Depending on the goal of the lesson, teachers may use learning tasks, practice tasks or assessment tasks. A learning task (Mok, 2004) is an example the teacher uses to demonstrate to students a new concept or skill or engage students to develop a concept or skill. Practice tasks are tasks used during the lesson to either illuminate the concept or demonstrate the skill further and tasks the teacher asks students to work through during the lesson either in groups or individually, or during out of class time (Kaur, 2010). Assessment tasks are tasks used to assess the performance of the students (Kaur, 2010).

Mathematical tasks can be examined from a variety of perspectives including the demands of the tasks and the presentation of the tasks. However, it is not always possible to subject all the tasks to the same type of analysis. As learning tasks, taken from textbooks or other sources, are set up for specific goals of instruction during the instructional cycle, these tasks cannot be treated in the same vein as review, practice and assessment tasks because the corresponding classroom discourse has a lot to do with how the students engage with it. Drawing on the framework for the analysis of learning task lesson events proposed by Mok and Kaur (2006), the purposes of learning tasks are differentiated by the following three levels of learning process, namely:

Level 1: introducing new concepts and skills;
Level 2: making connections between new and old concepts or skills;
Level 3: introducing knowledge or information beyond the scope of the curriculum requirement or textbook.

Stein and Smith's (1998) task analysis guide provides a way to establish the cognitive demands of the tasks used by the teachers in their classrooms. A brief outline of Stein and Smith's guide with adaptations made by the authors for the analysis of data presented in this chapter is shown in Table 1.

Table 1
Levels of cognitive demand

Levels of cognitive demand	Characteristics of tasks
Level 0 – [Very Low] Memorisation tasks	Reproduction of facts, rules, formulae No explanations required
Level 1 - [Low] Procedural tasks without connections	Algorithmic in nature Focussed on producing correct answers Typical textbook word - problems No explanations required
Level 2 [High] Procedural tasks with connections	Algorithmic in nature Has a meaningful / "real-world' context Explanations required
Level 3 – [Very High] Problem Solving / Doing Mathematics	Non-algorithmic in nature, requires understanding and application of mathematical concepts Has a "real-world" context / a mathematical structure Explanations required

In this chapter, we examine mathematical tasks used by two competent teachers to facilitate the learning of vectors by their grade 10 students. In our study, these teachers are "competent", where competency is a composite measure of their students' performance at examinations, their performance in class in the eyes of their students and fellow mathematics teachers. The research questions that guide the examination of mathematical tasks are:

> *What are the types and the cognitive demands of the mathematical tasks used by two competent teachers in facilitating the learning of vectors by grade 10 students in their lessons? How did these teachers implement these tasks?*

2 The Study

The data reported in this chapter is part of a large project, presently underway, which examines how competent and experienced secondary school mathematics teachers enact the mathematics curriculum (Kaur, Tay, Toh, Leong, & Lee, 2017). The project has a video-segment involving 30 teachers and a survey segment involving 600 teachers. The data collection for the video segment of the project is based on the renowned complementary accounts methodology developed by Clarke (1998, 2001).

2.1 *Participants*

The two teacher participants who are the focus of this chapter are Teacher 1 [T1] and Teacher 2 [T2]. Teacher 1 [T1] is a male who has taught mathematics for the last 20 years and Teacher 2 [T2] is a female teacher who has also taught mathematics for the last 20 years. Both teachers are lead mathematics teachers in Singapore. A lead teacher is one who is nationally recognised for his or her teaching competency and is trusted with the charge of developing fellow teachers in the school and the nation. The profile of the students of T1 and T2 are distinct. The students of T1 are grade 10 students comprised 11 boys and 11 girls of mixed abilities. The students of T2 are also grade 10 students comprised 10 boys and 25 girls of mixed abilities. The mathematical ability of students in the class of T1 was slightly below average as they were from the 40th percentile of their cohort and those in the class of T2 were from the 50th percentile of their cohort.

2.2 *Data*

In order to understand how the two teachers used mathematical tasks in their classrooms, instructional observations and video recordings were used. For both teachers, sequences of their lessons were recorded according to the protocol developed for the Learner's Perspective Study in Singapore (Kaur, 2009). The teachers were also interviewed periodically when their lessons were being recorded. All the instructional materials the teachers used for their lessons were documented. These materials and the teacher interviews are the source of data explored in this chapter.

The topic of Vectors was taught by T1 in ten lessons (each of 55-minute duration) and by T2 in eight lessons (each of 60-minute duration). The sequence of the sub-topics taught in the each lesson by T1 and T2 is shown in Tables 2a and 2b respectively.

Table 2a
Sequence of sub-topics taught in each lesson by T1

Lessons	Sub-topics
1	Introduction to vectors Magnitude of vectors
2	Column vectors Zero vectors Equal vectors Opposite vectors
3	Relationships of vectors Scalar multiple of a vector Operations on vectors
4 – 5	Resultant vectors and vector equation
6 – 10	Drawing of vector addition and subtraction Geometrical problems involving vectors

Table 2b

Sequence of sub-topics taught in each lesson by T2

Lessons	Sub-topics
1	Introduction to vectors
2 & 3	Addition & subtraction of vectors
4	Scalar multiple of a vector
5	Magnitude of a vector
6	Position vectors Expression of a vector in terms of two vectors
7 & 8	Geometrical problems involving vectors

2.3 *Data analysis*

The analysis of the data is guided by the following questions:

a) What were the types of mathematical tasks the teachers used? What were the cognitive levels of demand of the tasks? What guided the teachers in their choices of mathematical tasks for specific purposes?

b) How did the teachers enact the tasks?

The instructional materials used by T1 and T2 were compiled and systematically examined. The source of the tasks differed, as T2 used a collection of tasks that were developed over a period of time, while T1 used and adapted tasks from the textbook, in addition to tasks that were developed to supplement the achievement of lesson goals. When implementing the mathematical tasks, both teachers used pair discussion, teacher-student dialogue, and individual practice.

Counting of the mathematical tasks was conducted. When counting the mathematical tasks, even though there were different methods of presenting the mathematical tasks and engaging the students in the tasks (individual problem-solving or group discussion or teacher-student discussion), each mathematical task was counted as one when it was first presented by the teacher. If the teacher repeatedly used dialogue to guide the students in problem solving or in comprehending the same task, but the explored or learned mathematical concept was the same, it was still counted as one task.

To code the mathematical tasks, the authors constructed a classification system, adapted from Stein and Smith's (1998) and Mason and Johnston-Wilder's (2006), to place the tasks in each lesson into the six types and three levels of cognitive demands, as shown in Tables 3a and 3b. The six types are (A) Warming Up; (B) Building Skills; (C) Developing Understanding; (D) Practising Fluency; (E) Assessing Understanding; and (F) Making Connections. The tasks are also classified according to the level of cognitive demands, adapted from Mok and Kaur (2006), required when solving them: (1) Tasks that are algorithmic in nature and focus on producing correct answers; (2) Tasks that are algorithmic in nature but require explanations; and (3) Tasks that are non-algorithmic in nature and require application of mathematical concepts with explanations. Prior to the counting and classification of the mathematical tasks, the authors first clarified the definition of a task. Adopting the definition by Stein, Grover and Henningsen (1996), a mathematical task is defined as a set of problems or a single complex problem that focuses students' attention on a particular mathematical idea. The authors also discussed the classification standards, and then individually carried out classification based on transcripts of the recorded lessons of the two case teachers. Next, for inter-coder reliability, they discussed any different counting and classification results and modified the initial results to arrive at consensus.

During the 10 periods of observation, T1 administered a total of 47 mathematical tasks, of which 6 (12.8%) were for warming up, 5 (10.6%) for building skills, 9 (19.1%) for developing understanding, 18 (38.3%) for practising fluency, 6 (12.8%) for assessing understanding, and 3 (6.4%) for making connections; and of which 28 (59.6%) were of low cognitive demand and 19 (40.4%) of high cognitive demand. During the 8 periods of observation, T2 administered a total of 53 mathematical tasks, of which 2 (4.3%) were for warming up, 15 (31.9%) for building skills, 18 (38.3%) for developing understanding, 7 (14.9%) for practising fluency, 4

(8.5%) for assessing understanding, and 7 (14.9%) for making connections; and of which 30 (56.6%) were of low cognitive demand, 17 (32.1%) of high cognitive demand, and 6 (11.3%) of very high cognitive demand. The statistics and different types of mathematical tasks used by T1 and T2 are shown in Tables 3a and 3b respectively.

Table 3a

Number of mathematical tasked classified by types and by cognitive demands used by T1

	Types of mathematical tasks						Cognitive demands
	(A)	(B)	(C)	(D)	(E)	(F)	
Lesson 1	1	1					low
							high
							very high
Lesson 2	1		1	2			low
			4	1	1	2	high
							very high
Lesson 3				6			low
							high
							very high
Lesson 4		2	1	1	1		low
						1	high
							very high
Lesson 5	2			1			low
							high
							very high
Lesson 6	1			1			low
			1	2			high
							very high
Lesson 7		1	1	2			low
							high
							very high
Lesson 8							low
			1	1	1		high
							very high
Lesson 9	1				1		low
		1		1			high
							very high
Lesson 10							low
					2		high
							very high

Legend: Types of mathematical tasks – (A) Warming up; (B) Building skills; (C) Developing understanding; (D) Practising fluency; (E) Assessing understanding; (F) Making connections

Table 3b

Number of mathematical tasked classified by types and by cognitive demands used by T2

| | Types of mathematical tasks | | | | | | Cognitive demands |
	(A)	(B)	(C)	(D)	(E)	(F)	
Lesson 1		3					low
			1	1			high
							very high
Lesson 2	1	1	3				low
			1				high
							very high
Lesson 3		3	5	1			low
			1		1		high
							very high
Lesson 4	1	3	1				low
			1	1			high
					1		very high
Lesson 5		3	2				low
			1	1		1	high
							very high
Lesson 6		2	1				low
				1			high
						2	very high
Lesson 7							low
				1	1		high
						1	very high
Lesson 8							low
			1	1	1	1	high
						2	very high

Legend: As in table 3a

Figure 1 shows the analytical lens, adapted from The Teaching for Robust Understanding framework proposed by Schoenfeld, Floden, and the Algebra teaching Study and Mathematics Assessment Project (2014), that was crafted to examine lessons of the two teachers in Kaur, Wong, and Toh (2017).

Aspect of teacher's instructions	Indicators
Were important ideas in the lesson connected with those in past and future lessons?	Did the teacher connect the important idea/s in the lesson to what students already know? Did the teacher relate concepts to each other – not just in a single lesson, but also across lessons and units in past and future?
How were math procedures in the lesson justified and connected with important ideas?	How did the teacher develop mathematical knowledge in the class? (Telling and showing / developing concepts through student activities / through systematic logical steps) Did the teacher identify the important ideas behind concepts and procedures? Did the teacher highlight connections between skills and concepts?

Figure 1. Analytical lens to code teacher's instructions

The coding was done in the following manner. For each teacher, a lesson was segmented into episodes. An episode was delineated by the beginning and end of a mathematical task. Each episode was next scanned for indicators of the aspects of teacher's instructions. When a disagreement arose, the authors discussed their differences and arrived at consensus, either agreeing on the presence of the indicator or dismissing it.

3 Findings and Discussion

In this section we present the findings according to the questions that guided the analysis of the data.

3.1 *Mathematical tasks used by the teachers*

There were similarities and differences in the tasks used by the two teachers. To achieve the goals of their lessons, they had used similar types of tasks. However, to match the interest and ability of their students their tasks differed in context and cognitive demands. As shown in Tables 3a and 3b, the types of mathematical tasks implemented by T1 focused on practising fluency and developing understanding, while those by T2 focused on developing understanding and building skills. Just over a third of the tasks implemented by T1 is of high cognitive demand and the rest are of low, while almost half of the tasks implemented by T2 are of high or very high cognitive demands.

T1 believed in mastery by practise but not before a concept is first understood properly. Given the profile of the students, the teacher also placed emphasis on assessing understanding so as to gain feedback on students' learning difficulty in order to adjust the pace of lessons accordingly and provide timely and necessary support. T2 believed in the need to develop understanding and to build skills for the topic. Although tasks for assessing understanding seemed few, tasks that allow practising for fluency during the lesson had served to give the teacher sufficient feedback on students' learning progress. Given the profile of the students, the teacher also placed emphasis on making connections so that students were able to recognize and understand how mathematical concepts interconnect and build on one another to construct their own knowledge.

The choice of the cognitive demands of the tasks implemented by T1 was also based on the profile of the students. As the majority of the class were low progress learners, the tasks were chosen to motivate them to persevere at learning this new topic and to suitably challenge them so as to create a flow advocated by Csikszentmihalyi (Shernoff, Csikszentmihalyi, Schneider, & Shernoff, 2014). Similarly, based on the profile of the students, the choice of the cognitive demands of the tasks implemented by T2 was due to the belief that student learning is greatest in classrooms where the tasks consistently encourage high-level thinking and reasoning (Boaler & Staples, 2008).

Figures 2a to 2e show examples of the types of mathematical tasks used in the two teachers' lessons.

Draw the following: a) From a Point A, move 3 cm.
                          b) From a Point A, move 3 cm due North.

What is the difference in the instructions for (a) and (b) above?   A

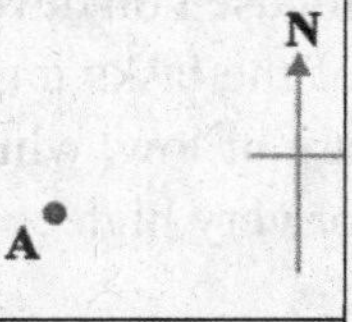

Purpose of task: To introduce the concept of a vector as having both a magnitude and a direction in contrast to a scalar quantity which has only magnitude.

Figure 2a. Type (A): Warming up

Draw the vector $p + q$ in each of the following.

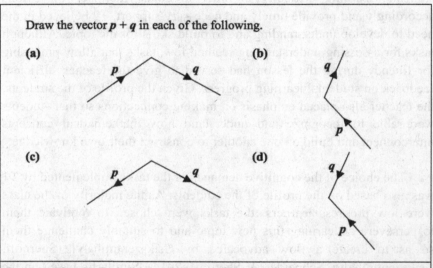

Purpose of task: To build the skills to add vectors, using Triangle Law of Vector Addition and verifying with Parallelogram Law, or vice versa.

Figure 2b. Type (B): Building skills

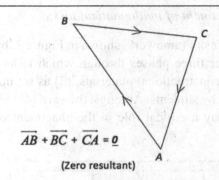

$$\overrightarrow{AB} + \overrightarrow{BC} + \overrightarrow{CA} = \underline{0}$$

(Zero resultant)

If you run from A to B, then to C, and back to A, what is your resultant displacement?

Purpose of task: To develop conceptual understanding of a zero vector (that it is not a scalar zero) and to use the concept to form equations for problem solving of unknown vector later.

Figure 2c. Type (C): Developing understanding

If a and b are two non-zero and non-parallel vectors, state whether the given statements are true.

(a) $|a+b| = |a| + |b|$

(b) $|a-b| = |a| - |b|$

(c) $|3a| = 3|a|$

(d) $|-3a| = -3|a|$

(e) $|-3a| = 3|a|$

Purpose of task: To assess the understanding of magnitude of a vector and of a resultant vector.

Figure 2d. Type (E): Assessing understanding

The coordinates of P, Q and S are $(1, 2)$, $(7, 3)$ and $(4, 7)$ respectively.
Using vector method, find the coordinates of R if $PQRS$ is a parallelogram.
How is this method different from what you have learnt in Coordinate Geometry?

Purpose of task: To make comparison between the two methods and make connections between the two topics.

Figure 2e. Type (F): Making connections

3.2 *Teachers' enactment of mathematical tasks*

The Mathematics Task Framework, shown in Figure 3, by Stein and Smith (1998) distinguishes three phases through which tasks pass: (1) as they appear in curricular/instructional materials; (2) as set up by teachers; and (3) as implemented by students. Amongst the variables within the different phases, teachers play a critical role in the enactment of a mathematical task.

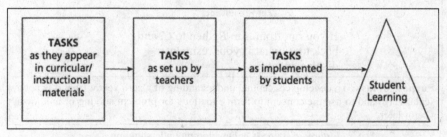

Figure 3. The Mathematics Task Framework (Stein & Smith, 1998, p.270)

The level of cognitive demand of a task can be altered depending on the enactment by the teacher. This means that a task could be labelled of low cognitive demand but with effective questioning and probing by the teacher, it can become a high cognitive demand task. Of course, similarly, if the teacher limits students' thinking, a task can be turned from a high into a low cognitive demand one.

The following show sample excerpts from a particular lesson of each of the two teachers that were evidence of the presence of the indicators under the analytical lens that demonstrated how the two teachers (1) connected important ideas in the lesson with those in past and future lessons; and (2) connected math procedures in the lesson with important ideas, in their efforts to actively engage their students with mathematical ideas during lessons, even when the mathematical tasks were of low cognitive demands.

<u>Were important ideas in the lesson connected with those in past and future lessons?</u>

Did the teacher connect the important idea/s in the lesson to what students already know?

> T2: (05:05) How do you represent your vectors when you do Science?

Did the teacher relate concepts to each other – not just in a single lesson, but also across lessons and units in past and future?

> T1: (42:21) Many quantities have only magnitude... you are all familiar with that in the primary school. When you come to secondary school, you started learning in physics, ... These are the various quantities that you are familiar with.

How were math procedures in the lesson justified and connected with important ideas?

How did the teacher develop mathematical knowledge in the class?

> T1: (08:37) Now, what did you observe about these four vectors? How are they different and how are they the same? (09:14) What other observations did you observe? (10:08) What do you notice about OA and OC?

> T2: (10:43) [Do] You notice I use the word displacement? Why do I use the word displacement?

Did the teacher identify the important ideas behind concepts and procedures?

> T1: (31:10) Can you see that there are two pairs of vectors which are related to each other? (35:16) Two lines are parallel but do they need to be in the same direction?

> T2: (11:15) Why is it useful that we represent a vector using x/y like this [column vector]?

Did the teacher highlight connections between skills and concepts?

> T2: (15:00) If your vector is not represented by a column vector, then how do you find the magnitude? ... And you will use all kinds of knowledge that you have to find length. (16:49) Look at the diagram and ask yourself, what do you know? What are the

concepts, what are the skills your already have? What can you use to find …

4 Conclusion

Teachers use mathematical tasks to engage students in activities and facilitate learning. Individual teachers arrange instruction very differently, depending on what they are teaching, and students respond to instruction very differently, depending on the structure and demands shaped by tasks enacted in the classroom. In this chapter, we examined the types of mathematical tasks used and how these tasks were enacted by two competent teachers to facilitate the learning of vectors by grade 10 students.

The data presented in this chapter showed that there were similarities and differences in the tasks they used. To match the goals of their lessons, they used similar types of tasks. However, to match the interest and ability of their students their tasks differed in context and cognitive demands. The commonalities between the mathematical tasks used by the two teachers were (1) the tasks were designed to actively engage students in mathematical thinking; (2) the tasks took into account students' previous knowledge and experiences; and (3) appropriate teacher facilitation through scaffolding and questioning supported students' understanding of the mathematical concepts involved in the tasks.

The nature of tasks often changes as they pass from one phase to another. In other words, a task that appears in the curricular or instructional materials may not always be identical to that enacted by the teacher; in turn, it is not always exactly the same that the students actually do (Stein & Smith, 1998). The data presented in this chapter also brought out the importance of the role of teachers as they enact the mathematical tasks in order to support meaningful connections between procedures, concepts and contexts, and also to provide opportunities for students' engagement in key practices such as reasoning and problem solving, so as to be aligned with the goals stipulated by the school mathematics curriculum.

The data and findings presented in this chapter are from the teachers' perspective. To gain a more holistic insight to how mathematical tasks are enacted in the school mathematics curriculum, we need to look at it from the learners' perspective and study the circumstances that allow students to engage in and spend time on, in other words, to examine the opportunities to learn mathematics afforded by tasks teachers enacted in their classrooms (National Research Council, 2001; Watson, 2007).

Acknowledgement

This chapter is based on the Programmatic Project: A Study of the Enacted School Mathematics Curriculum (OER 31/15 BK) at the National Institute of Education, Nanyang Technological University, Singapore.

References

Boaler, J., & Staples, M. (2008). Creating mathematical futures through an equitable teaching approach: The case of Railside School. *Teachers College Record, 110*(3), 608-645.

Christiansen, B., & Walther, G. (1986). Task and activity. In B. Christiansen, G. Howson, & M. Otte (Eds.), *Perspectives on mathematics education* (pp. 243-307). Dordrecht, Reidel.

Clarke, D. J. (1998). Studying the classroom negotiation of meaning: Complementary accounts methodology. Chapter 7 in A. Teppo (Ed.) *Qualitative research methods in mathematics education,* monograph number 9 of the *Journal for Research in Mathematics Education,* Reston, VA: NCTM, 98-111.

Clarke, D. J. (Ed.) (2001). *Perspectives on practice and meaning in mathematics and science classrooms.* Dordrecht, Netherlands: Kluwer Academic Press.

Doyle, W. (1983). Academic work. *Review of Educational Research, 53*(2), 159-199.

Hiebert, J., & Wearne, D. (1993). Instructional tasks, classroom discourse, and students' learning in second-grade arithmetic. *American Educational Research Journal, 30*(2), 393-425.

Kaur, B. (2009). Characteristics of good mathematics teaching in Singapore grade 8 classrooms: a juxtaposition of teachers' practice and students' perception. *ZDM Mathematics Education, 41*(3), 333-347.

Kaur, B. (2010). A study on mathematical tasks from three classrooms in Singapore. In Y. Shimizu, B. Kaur, R. Huang, & D. Clarke (Eds.), *Mathematical tasks in classrooms around the world* (pp. 15-33). Sense Publishers.

Kaur, B., Tay, E.G., Toh, T.L., Leong, Y.H., & Lee, N.H. (2017). A study of school mathematics curriculum enacted by competent teachers in Singapore secondary schools. *Mathematics Education Research Journal. 30*(1), 103-116. https://doi.org/10.1007/s13394-017-0205-7

Kaur, B., Wong, L.F., & Toh, T.L. (2017). A framework to examine the mathematics in lessons of competent mathematics teachers in Singapore. In B. Kaur, W.K. Ho, T.L. Toh, & B.H. Choy (Eds.). *Proceedings of the 41st Conference of the International Group for the Psychology of Mathematics Education, Vol. 3*, pp. 41-48. Singapore: PME.

Mason, J., & Johnston-Wilder, S. (2006). *Designing and using mathematical tasks*. St Albans, UK: Tarquin Publications.

Mok, I.A.C. (2004). *Learning tasks*. Paper presented at the Annual Meeting of the American Educational Research Association, San Diego, April 12-16, 2004.

Mok, I.A.C., & Kaur, B. (2006). "Learning Task" lesson events. In D. Clarke, J. Emanuelsson, E. Jablonka, & I.A.C. Mok (Eds.), *Making connections: Comparing mathematics classrooms around the world* (pp. 147-164). Rotterdam, The Netherlands: Sense Publishers.

National Research Council. (2001). Adding it up: Helping children learn mathematics. *Mathematics Learning Study Committee, Center for Education, Division of Behavioral and Social Sciences and Education*. Washington, DC: National Academy Press.

Schoenfeld, A.H., Floden, R.E., & the Algebra Teaching Study and Mathematics Assessment Project (2014). An introduction to the TRU Math Dimensions. Berkeley, CA & E. Lansing, MI: Graduate School of Education, University of California, Berkeley & College of Education, Michigan State University. Retrieved from: http://ats.berkeley.edu/tools.html and/or http://map.mathshell.org/materials/pd.php

Shernoff, D. J., Csikszentmihalyi, M., Schneider, B., & Shernoff, E. S. (2014). Student engagement in high school classrooms from the perspective of flow theory. In *Applications of Flow in Human Development and Education* (pp. 475-494). Springer Netherlands.

Shimizu, Y., Kaur, B., Huang, R., & Clarke, D. (2010). The role of mathematical tasks in different cultures. In Y. Shimizu, B. Kaur, R. Huang, & D. Clarke (Eds.), *Mathematical tasks in classrooms around the world* (pp. 1-14). Sense Publishers.

Stein, M. K., Grover, B. W., & Henningsen, M. (1996). Building student capacity for mathematical thinking and reasoning: An analysis of mathematical tasks used in reform classrooms. *American Educational Research Journal, 33*(2), 455-488.

Stein, M. K., & Smith, M.S. (1998). Mathematical tasks as a framework for reflection: From research to practice. *Mathematics Teaching in the Middle School, 3*(4), 268-275.

Watson, A. (2007). The nature of participation afforded by tasks, questions and prompts in mathematics classrooms. *Research in Mathematics Education, 9*(1), 111-126. https://doi.org/10.1080/14794800008520174

Chapter 5

Use of Comics and Its Adaptation in the Mathematics Classroom

TOH Tin Lam   CHAN Chun Ming Eric   CHENG Lu Pien
LIM Kam Ming   LIM Lee Hean

We have developed an alternative teaching package for selected chapters of Lower Secondary mathematics using comics and storytelling under the research project entitled MAGICAL (MAthematics is Great: I Can And Like) on using comics in mathematics education. This chapter reports how the participating teachers from one research school adopted and adapted this alternative comics teaching package for instruction in their mathematics classes, and the innovative practices that they used in these lessons. We also report our long-term plan in scaling up this approach of using comics to more schools and teachers through the use of appropriate semi-structured mentoring framework.

1 Introduction

The use of comics in mathematics education has received the attention of educators throughout the world and, in particular, in numerous schools in Singapore. It is generally recognized that low attaining students in mathematics, who usually lack interest in the subject, are likely to be attracted to comics and cartoons. Thus, the use of comics in education has the potential to capture the attention and interest of students at the school-going age. In other words, comics, if used appropriately in education settings, has the ability to address the affective aspect of learning a subject, in particular, mathematics (Şengül & Üner, 2010; Tilley, 2008; Toh, 2009).

In primary schools, teachers have used comics and stories in the mathematics classroom. Although formal primary school mathematics textbook presentation is deemed to be laden with graphics and well-sequenced illustrations to help bring mathematics ideas across to children, they are still a staid form of instructional material that appeals to a certain degree of seriousness in disposition at the hands of teachers who use them to teach. The use of stories and comics thus shifts the instructional approach for children to learn mathematics with a softer touch so that learning appeals both cognitively and affectively to children's enjoyment, interest and excitement. Cognition, emotion and motivation are key factors that influence the way children feel about mathematics (Hannula, 2006; Shmakov & Hannula, 2010). In a similar light, it is deemed that secondary school students could also benefit from such an instructional approach. The use of comics cartoons can improve retention of students' understanding of complex ideas (Wylie & Neeley, 2016).

The objective of impacting the teaching and learning of mathematics cognitively and affectively through the use of stories and comics forms the basis of this chapter through a research project for lower secondary mathematics. Toh, Cheng, Jiang, and Lim (2016) reported the details and the principles of developing comics package as an alternative teaching package for lower secondary mathematics under the research project MAGICAL (MAthematics is Great: I Can And Like). The comics alternative teaching packages developed under MAGICAL were improved and several more sets of packages were developed under the research project SUPER-MAGICAL (Scaling UP the Educational Research MAGICAL).

The comics alternative teaching package on each lower secondary mathematics topic includes: (1) a complete set of comics that explicates all the mathematical concepts within the curriculum through incidents (mainly humorous, funny or even ridiculous) by the fictitious characters in the comics; (2) a full set of tiered practice questions with varying levels of difficulties; and (3) a corresponding set of lesson plans proposed by the researchers (hereafter shall be referred as "we") on how storytelling could be infused in reading the comics, and the rationale of the approaches proposed by us.

These alternative teaching packages were given to the SUPER-MAGICAL research schools and were subsequently used by the research schools in their usual mathematics instruction as a "replacement" of their usual teaching textbook resources. We were keen to study (a) the impact of the use of comics on students' motivation in mathematics, mathematical self-concept and performance in mathematics achievement test; and (b) how the teachers adapted the package in the lesson implementation.

This chapter reports how the participating teachers in one of the SUPER-MAGICAL research schools adapted the teaching package on Lower Secondary Percentage in their lessons. We also present a brief discussion of our long-term plan in scaling up this approach of using comics to a wider circle of mathematics teachers in Singapore.

2 Literature Review

In this chapter, we shall use the term *comics* to refer to an approach of conveying ideas (which could be stories, and which might not necessarily be realistic—and, more often than not, imaginary, humorous and possibly hilarious stories) through a sequence of visual images—usually in the form of cartoons (Toh, Cheng, Jiang, & Lim, 2016; Toh, Cheng, Ho, Jiang, & Lim, 2017).

Literature abounds with reasons why comics should be used for education, although in the old days, comics were usually seen as the "enemy" of schools. There are some inherent features of comic that make it possible—even sensible—as a pedagogical tool in education. Firstly, the abundant use of visual arts in comics facilitates the reader to create their own understanding of the information presented in textual forms.

According to McVicker (2007), visual arts have the ability to guide students' understanding of spatial relationship within the context.

Secondly, the integration of comics and mathematics makes the latter intuitive and impressive to the students, so that students are more likely to have successful learning and understanding when they are presented in a way that is meaningful and interesting to them (Price & Lennon, 2009).

In addition, the humor that accompanies comics usually leaves an impression upon readers on the message it conveys. The seminal work on humor in education by Kaplan and Pascoe (1977) asserts that humor can create the long-term retention of concepts. Subsequent studies have also shown that humor encourages student retention of information and creates an environment conducive for learning (Wanzer, Frymier, & Irwin, 2010; Segrist & Jupp, 2015).

Several studies related to mathematics education have shown that comics and cartoons increase students' motivation and interest in learning mathematics (Cho, 2012). Cartoons and comics are even impactful on teachers! The study by Cho, Osborne, and Sanders (2015) on pre-service teachers shows that the latter's level of engagement and enjoyment in mathematics increased when comics and cartoons were used in the mathematics courses. Another important point to note is that cartoons and comics have been reported to have the ability to reduce students' anxiety level about the subject (Şengül & Dereli, 2010).

Comics can tap on the use of storylines that are attractive to students. The use of stories, being the main feature of comics, has been shown to be an effective tool in education. According to Sipe (2002), there are five ways that stories can be used to get students to be engaged "with words and physical actions" (p. 477). The teacher can engage the students in (i) dramatizing the story, both verbally and non-verbally; (ii) talking back to the story or characters within the story; (iii) critiquing or controlling the plot where the students are given the opportunity to suggest alternative plots, characters, settings, and so on; (iv) inserting themselves or their friends in the story where they get themselves or their friends to assume the role of story in role-playing; and (v) taking over the text and manipulate the text for their own purposes.

In other words, students may use the story as a launching pad to express their creativity. The common characteristic in all these five types of response is that students who made these responses view stories as "invitations to participate or perform" (Sipe, 2002, p. 479) and were not mere passive learners in the classroom.

3 Implementation of the Comics Teaching Package

In both research projects MAGICAL and its subsequent scaling-up project SUPER-MAGICAL, we selected five mathematics topics in the lower secondary mathematics curriculum (Percentage, Statistics, Application of Mathematics, Data Analysis and Introduction to Probability) and developed a comics teaching package for each topic.

We proposed two different modes of instruction that the participating teachers can implement their comics lessons: (1) the use of a hardcopy version of the teaching package; and (2) an online version (samples of the comic strips are available at the website http://math.nie.edu.sg/magical). The teachers could decide on one of the versions (or a blend of the two versions) that best suited their needs and the constraints of the school environment.

Each set of comics teaching package was designed to replace the existing teaching resource (that is, textbooks and the accompanying resources provided by the textbook publishers and the resource provided by the Ministry of Education) for that particular topic. The teachers in the participating schools were requested to use the comics teaching package as the substitute resource for the selected topics mentioned above. One of the research goals of the project SUPER-MAGICAL is to study the impact of using comics and storytelling on students' motivation, mathematics self-concept and performance in the mathematics achievement test.

Another research goal of the project SUPER-MAGICAL is to identify how school teachers adopted or adapted the comics teaching package for their teaching in the mathematics classroom. Through this research, we hope to identify a set of teaching approaches that is used in the mathematics classroom associated with the use of comics. In order to achieve this, all the mathematics lessons taught using the comics teaching package were video-recorded and analyzed.

This chapter reports our observation in one particular Singapore mainstream school participating in SUPER-MAGICAL (hereafter, shall be referred as "the research school") in implementing the first topic on Percentage. The research school adapted our comics teaching package for teaching one lower secondary Normal Technical mathematics class, (which we shall call Sec 1A). We report how the teachers adapted our existing package prior to the actual lesson implementation, and the prominent pedagogical practices that were observed during their lessons.

The class Sec 1A was co-taught by two teachers, Mr Henry and Ms Janice[1] throughout the year. In particular, both teachers took turns in teaching the selected portions of the topic Percentage. When one teacher was the main instructor for the lesson, the other teacher would assist the main instructor in monitoring students' progress in the class.

3.1 *Prior to the comics lessons*

Prior to the lessons using comics teaching package, the researchers and the two teachers held several meetings discussing the underlying ideas of using comics for teaching, and the details on the actual lesson implementation. The teachers were informed that they were given the option of either using the printed copies of the comics or the online version. We communicated to the teachers that one feature of the online version of the comics which the teachers could consider using was its

---

[1] To ensure confidentiality of the teachers, pseudonyms were used for the teachers and the name of the class.

interactive feature in the sense that students would be able to get immediate feedback on their performance in the practice questions.

The teachers decided to tap mainly on the use of hardcopy of the printed comics before they embarked on the teaching of Percentage. One main reason of consideration for choosing the printed version over the online version was the constraint on the availability of the computer laboratories. However, they also decided to use the online version of the comics package for a few selected lessons during which the computer laboratories were available.

The teachers modified the printed copies of comics by removing certain figures and leaving them as blanks for students to complete (see Figure 1 for a sample of the modified comic worksheet for students).

Figure 1. A Sample Page of the Comics Adapted by the Teachers

Instead of getting students to be passively listening to the story, the teachers adapted to the feature of getting students to co-construct the story, especially the mathematical concepts found in the story.

This is in alignment with Sipe (2002)'s perception of the pedagogical values of stories in education:

> Stories are understood not as fixed and rigid but as changeable texts, and the reader's role is not simply to understand but to actively control stories. We can change stories, resist them, critique them, even use them for our own purposes. These five types of response not only show children actively engaging with stories, they show children making stories their own (p. 479).

4 Observations during the Comics Classroom Lessons

4.1 *Setting the ground in the first lesson on Percentage*

Ms Janice started the first part of the first lesson on percentage with a quick recap of concepts related to fractions and decimals. This is because the concept of percentage builds on students' pre-requisite knowledge of fractions, decimals, and the conversion between them. She started the lesson by giving several real-life examples of where decimals, fractions and percentages are used in real world contexts as a means to seamlessly introduce the percentage notation.

Ms Janice selected familiar contexts to ensure students' association of the prior knowledge when she got students to recall the concept of fractions as part of a whole before moving on to what was to be covered. As the subsequent comics involved two fictitious characters Sam and Sarah observing an interesting event in the pizza shop, Ms Janice aptly used the context of pizzas to facilitate students' recall of the fundamental ideas of parts of a whole in a fraction.

Mr Henry continued with the second part of the first lesson by discussing real-life experiences involving GST and service charges, which are important in the local context, and stressed how these entities are usually stated in percentage rather than absolute amounts. Much time was spent by Mr Henry to elaborate on the importance of quoting these quantities in percentages.

As Mr Henry continued with narrating the story of the journey of the two fictitious characters (as proposed in the set of lesson outline proposed by the researchers), he was constantly emphasizing the related mathematical concepts, consciously chunking the concepts within the lesson and regularly consolidating each particular concept by engaging students with the practice questions accompanying the teaching package.

4.2 *Engaging students in the comics lessons*

Throughout the entire set of lessons on percentage (refer to the Appendix), the teachers gradually increased the level of student engagement in the lessons. In the first lesson, Mr Henry began his portion of the lesson by setting the scene on the use of percentages in a journey in a shopping centre. He began with telling the story proposed in the comics. In the next lesson, he got his students to play the role of the characters in the comics by reading the portion of the texts provided. At appropriate junctures, he interjected by inviting the entire class to discuss the related mathematical concepts that were revealed in the comic strips.

As the lessons progressed, both Mr Henry and Ms Janice involved more students to participate in the classroom discourse: in each lesson, two students were selected to role-play each of the two main characters in the comic strips. They were expected to complete the blanks that has been deliberately created by the teachers (an example is shown in Figure 1). Furthermore, the other students in the class were invited to ask questions related to the mathematical concepts revealed in the comic strips. To further engage the students, the students in role-play were also allowed to question the other students on the related mathematical concepts.

Here we observed that the teachers gradually reduced the amount of scaffolding during the lesson. Using the comics as a context, the teachers began with storytelling using the comics in teaching the related mathematical concepts, and moved from the main participant of the classroom discourse to the facilitators of the discourse. The students became the main participants in the classroom discourses, with teachers interjecting when new concepts need to be introduced to the students.

Research has found that students' perceptions of multiple types of motivation and emotion are key factors that contribute to their engagement and learning. Linnenbrink-Garcia, Patall, and Pekrun (2016) outlined five concepts that are influential in promoting adaptive student motivation and emotion: (a) support students' feelings of competence; (b) enhance autonomy; (c) use personally relevant and active tasks; (d) emphasize learning and de-emphasize social comparison; and (e) encourage feelings of belonging.

The gradual move from a teacher-led scaffold to a student-led participatory lessons is informed by the central tenets of autonomy, affiliation, and competence articulated by the Self-Determination Theory (Deci, Koestner, & Ryan, 2001; Irvine, 2017; Stipek, 2002). The Self-Determination Theory postulated that pedagogies that enhances students' sense of autonomy, affiliation and competency are more likely to successfully motivate students to persist on attempting and learning from tasks that they initially perceived to be difficult or uninteresting.

Autonomy is enhanced through the gradual release structure during which students gained increasing control over their learning. Using comics as a facilitative scaffold to increase students' competence in solving mathematics problems based on teachers-led lessons, the students moved to a greater level of autonomy in achieving mastery of the mathematical concepts without significant teacher intervention.

4.3 *Interpreting the context of the comics*

It was interesting to observe that the teacher (T) Ms Janice interpreted the mathematical concepts in the context of the comics with the students (S). This was evident in the fourth lesson in which the teacher Ms Janice interpreted the "meaning" of 100% in the context of a witch who declared she was 100% accurate in fortune-telling, before she built on that 100% is one whole.

T:	What does 100% accurate mean?
S:	Very accurate.
T:	How accurate?
S:	Very very accurate.
T:	Suppose she has 100 customers. How many of them will have accurate reading?
S:	100.
T:	What about 20 customers?
S:	20.

In the same lesson, Ms Janice created the impact of a percentage of more than 100 is greater than one whole. She used the same set of comic strips which showed the statement "Refund 150%".

She invited the students in perspective taking and created a scenario using numerical values:

T:	Suppose you pay $40 and you get 150% refund. Will you get more or less than the amount you pay?
S:	More.
T:	[By h]ow much?
S:	$40 + $20 … $60.
T:	Yes. 100% is $40 and 50% is $20 (used the whiteboard).

The teacher acknowledged the student's response (in splitting 150% into 100% + 50%) and continued to demonstrate the calculation of the actual amount by direct calculation. The teacher continued with consolidation by giving the students three more examples to calculate: 120%, 200%, 120.5% of $40.

4.4 *Use of unrealistic examples to impress upon students*

Our comics package includes the use of "lame jokes" and exaggerated cases, which are far from realistic problems of the world. This is a different paradigm from the Ministry of Education's emphasis on the use of real-world (which implies realistic) problems in mathematics education. Along the line of thought of communication theory, we believe that the use of jokes and exaggerated examples could further impress upon students the associated mathematical concepts attached to the context.

As Mr Henry discussed the second set of comics on "whale watching", as the fictitious character Sarah highlighted that she would spend 50% of her monthly salary of $3000 on a boat ride for whale watching, Mr Henry stressed that he would not spend such a large part of his salary for a boat trip.

> T: I would not spend so much portion of my salary on a boat ride. What do you think?
> S: (discussing among themselves).

This is an instance where the teacher, instead of shunning from such an unrealistic event, makes use of this opportunity to raise the awareness of the real world among his students.

In another instance in the fifth lesson, Mr Henry discussed the concept of percentage increase or decrease. Noticing that the students had already acquired the concepts and skills in computing percentage changes, he created, using the class context, a problem that has an extremely unrealistic answer.

He used one student's example of coming to school by taxi when she was late. It cost her about $8. Previously, she had travelled by bus which had cost her $0.80. Mr Henry used this episode as an example for students to calculate the percentage increase in the student's expenditure if she had decided to subsequently travel by taxi to school. The answer to this question was enormous in terms of percentage. This is another instance when the teacher created a contextual problem with an extremely exaggerated outcome, to impress upon the students.

4.5 *Building contextual knowledge in students*

We observed that comics provided the teachers with opportunity to discuss real-world context knowledge with their students. This was already reported in Toh et al. (2017) on how another group of teachers exploited the opportunities to develop students' 21st century competencies using comics.

The recorded instances in this pair of teachers we observed in the first three sets of comics are discussed below:

Comics Set 1 (Let's Go Shopping). The story of this set of comics has its setting in a shopping mall during the Great Singapore Sales period. The teacher used the opportunity to discuss the reasons why Singaporeans like to make purchases during the Great Singapore Sales period, how discounts are always shown in the real world, and that purchases made in Singapore usually entails Goods and Services Taxes and service charges.

Comics Set 2 (Go for Whale Watching in Singapore). The setting of this story is at a particular harbour in Singapore. The teacher alerted the students that whales are usually not seen in Singapore, although there could be isolated instances that whales are found dead on the Singapore seaside.

Comics Set 3 (Tell Your Future). This story sets the context of a fortune-teller store along the Singapore cultural heritage site. The teacher Ms Janice took the opportunity to discuss the Singapore cultural heritage sites in Singapore that belong to the various races in Singapore.

5 Discussion

This chapter reports our observation of two teachers co-teaching one Secondary One class in a Singapore mainstream school. The researchers provided the teachers with a set of proposed lesson outlines accompanying the comic strips. The teachers adapted the package to meet their learners' needs. Although the researchers provided the teachers with the full story that could be used with each set of comics, the teachers chose to engage the students in contributing towards the classroom mathematics discourse by gradually introducing role-play and student questioning. The process of engaging students was also done gradually, starting from teacher storytelling to student reading comic passages and eventually to student role-play.

The teachers also seized opportunities in the context of the comic strips to introduce to students real world knowledge, and to get students to do role-playing (as to view the incident from the view of the fictitious characters in the comics). Wery and Thomson (2013) found that students who can see the links between a lesson task and the real world are likely to be more motivated to learn to solve the problem posed. Using comics with relevant real life applications and authentic tasks enhances the students' motivation and helps them to be effective learners (Chua, 2014).

In line with the assumptions of the Self-Determination Theory (Deci et al., 2001), class content usefulness and relevance to real life context is associated with harmonious passion (Ruiz-Alfonso & León, 2017). Harmonious passion, which is defined as the enjoyment of completing the associated tasks, was also positively correlated with mathematics achievement (Ruiz-Alfonso & León, 2017).

Role-play helps to heighten the sense of relevance and applicability of the class content to the real world. This in turn motivate students to take a greater interest in the task and to persist on working on the task. In addition, Williams (2014) found that role-play is itself an effective tool to facilitate mathematical learning amongst primary school students. Role-play helps to strengthen students' engagement as participants in a community of learners.

Throughout the lessons, the teachers were clearly very mindful of their instructional goals, and at the same time, they were fully cognizant of the need to adapt the comics lessons to their students' ability in learning and even swapped the sequence of the sets of comics provided for them. The teachers also value-added to the comics package by creating their own numerical examples with extraordinarily large numbers to impress upon their students the various mathematical concepts, in addition to developing their procedural fluency.

What was also observed was the use of comics provided the teachers with a platform to discuss other contextual knowledge (for example, knowledge of the country) provided by the platform. In an earlier study by Toh et al. (2017), we also observed that teachers incorporated features that enhance students' development of 21st century skills by riding on the affordances of the comics platform.

6 The Next Step: How to Scale Up

The SUPER-MAGICAL research team has implemented the use of comics in the mathematics classroom in the initial research schools, and we have observed many good pedagogical practices among the initial team of participating teachers. Our next step in this project is to produce more teachers who are able to use this approach in teaching mathematics. We believe that this is a meaningful challenge. In this study, we adopted a semi-structured mentoring framework to facilitate both mentoring multiplier and individual mentor-protégé learning relationships. We began our research by conducting briefings for the potential participants, which serve to ascertain their interest and commitment to participate in our study. Upon their commitment, the researchers discussed in greater depth with the teacher participants the issues of lesson implementation, the pedagogical principles and the real world constraints.

The following initial components of the mentoring framework are, namely, Identify Potential Mentors/Protégés to Establish/Consolidate Connections (through NIE initiated communication prior to School Implementation), and Review/Assess Readiness of Mentors to influence protégés positively. Video recordings provide useful means to facilitate the identification and transfer of strategies. Subsequent concurrent components of the mentoring framework incorporate team sharing to facilitate Mentors-Protégés Interactions/Exchanges and enhance mentoring multiplier. We believe that this is an essential step in the process of scaling up our approach to more Singapore schools.

7 Conclusion

The use of stories and comics makes for a non-threatening and engaging learning environment for young learners. With mathematics embedded as part of the features in the design of the story or comics plot, there is potential to address both cognitively and affectively the learning of mathematics through such a medium.

Acknowledgement

The authors would like to thank all the teachers and students who participated in this study. This research project is funded by the Singapore National Institute of Education under AFR 4/16 TTL.

References

Cho, H. (2012). *The use of cartoons as teaching tool in middle school mathematics.* Seoul: ProQuest, UMI Dissertation Publishing.

Cho, H., Osborne, C., & Sanders, T. (2015). Classroom experience about cartooning as assessment in pre-service mathematics content course. *Journal of Mathematics Education at Teachers College, 6,* 45-53.

Chua, Y. P. (2014). The effects of humor cartoons in a series of bestselling academic books. *Humor,* 27(3), 499-520.

Deci, E., Koestner, R., & Ryan, R. (2001). Extrinsic rewards and intrinsic motivation in education: Reconsidered once again. *Review of Educational Research, 71,* 1-27.

Hannula, M. S. (2006). Affect in Mathematical Thinking and learning. In J. Maaß, & W. Schlöglmann (Eds.), *New mathematics education research and practice* (pp. 209-232). Rotterdam: Sense.

Irvine, J. (2017). Problem posing in consumer mathematics classes: Not just for future mathematicians. *The Mathematics Enthusiast, 14,* 387-412.

Kaplan, M., & Pascoe, C. (1977). Humorous lectures and humorous examples: Some effects upon comprehension and retention. *Journal of Educational Psychology, 69,* 61-65.

Linnenbrink-Garcia, L., Patall, E. A., & Pekrun, R. (2016). Adaptive Motivation and Emotion in Education. *Policy Insights from the Behavioral and Brain Sciences, 3,* 228-236.

McVicker, C.J. (2007). Comic strips as a text structure for learning to read. *The Reading Teacher, 61,* 85-88.

Price, R. R. & Lennon. C. (2009). *Using children's literature to teach mathematics.* Durham, NC: Quantile.

Ruiz-Alfonso, Z., & León, J. (2017). Passion for math: Relationships between teachers' emphasis on class contents usefulness, motivation and grades. *Contemporary Educational Psychology, 51,* 284-292.

Segrist, D. J., & Jupp, S. D. (2015). This class is a joke! Humor as a pedagogical tool in the teaching of psychology. *Psychology Teacher Network: American Psychological Association, 25,* 14-15.

Şengül, S., & Dereli, S. (2010). Does instruction of "integers" subject with cartoons effect students' mathematics anxiety? *Procedia – Social and Behavioral Sciences, 2,* 2176-2180.

Şengül, S., & Üner, İ. (2010). What is the impact of the teaching "Algebraic Expressions and Equations" topic with concept cartoons on the students' logical thinking abilities? *Procedia-Social and Behavioral Sciences, 2,* 5441-5445.

Shmakov, P., & Hannula, M. S. (2010). Humour as means to make mathematics enjoyable. *Proceedings of Congress of European Research in Mathematics Education, 6,* 144-153.

Sipe, L. R. (2002). Tallking back and taking over: Young children's expressive engagement during story book read-alouds. *The Reading Teacher, 55*(5), p. 476-483.

Stipek, D. (2002). Good instruction is motivating. In A. Wigfield, & J. Eccles (Eds.), *Development of achievement motivation* (pp. 309-332). San Diego, CA: Academic Press.

Tilley, C. L. (2008). Reading comics. *School Library Media Activities Monthly, 24*, 23-26.

Toh, T. L. (2009). Use of cartoons and comics to teach algebra in mathematics classrooms. In D. Martin, T. Fitzpatrick, R. Hunter, D. Itter, C. Lenard, T. Mills, & L, Milne (Eds.), *Mathematics of prime importance: MAV yearbook 2009* (pp. 230-239). Melbourne: The mathematical Association of Victoria.

Toh, T. L., Cheng, L. P., Jiang, H., & Lim, K. M. (2016). Use of comics and storytelling in teaching mathematics. In P. C. Toh, & B. Kaur (Eds.), *Developing 21st century competencies in the Mathematics classroom* (pp. 241-259). Singapore: World Scientific.

Toh, T. L., Cheng, L. P., Ho, S. Y., Jiang, H., & Lim, K. M. (2017). Use of comics to enhance students' learning for the development of the twenty-first century competencies in the mathematics classroom. *Asia Pacific Journal of Education, 37,* 437-452. https://doi.org/10.1080/02188791.2017.1339344

Wanzer, M. B., Frymier, A. B., & Irwin, J. (2010). An explanation of the relationship between instructor humor and student learning: Instructional humor processing theory. *Communication Education, 59,* 1-18.

Wery, J., & Thomson, M. M. (2013), Motivational strategies to enhance effective learning in teaching struggling students. *Support for Learning, 28,* 103-108. doi:10.1111/1467-9604.12027.

Williams, H. J. (2014). *The relevance of role play to the learning of mathematics in the primary classroom.* Doctoral dissertation, Roehampton University.

Wylie, C., & Neeley, K. (2016). Learning Out Loud (LOL): How comics can develop the communication and critical thinking abilities of engineering students. *Proceedings of the 2016 American Society for Engineering Education Annual Conference.* DOI: 10.18260/p.25542

Appendix

This appendix outlines the sets of comics that were designed by the researchers on the lower secondary topic Percentage.

Set	Title of Comics	Mathematical Content
1	Let's Go Shopping	• Recall meaning of percent • Convert percent to fraction • Convert percent to decimal
2	Go for Whale Watching in Singapore	• Convert fraction to percent • Convert decimal to percent • Find an amount given percent
3	Tell Your Future	• Calculation involving percent more than 100
4	Having Fun in Wonderland	• Express one quantity as a percentage of another
5	Discount!	• Calculate percentage discount
6	Office Politics	• Calculate an amount given its percentage
7	Getting Expensive	• Calculate percentage increase • Calculate percentage decrease

Chapter 6

Designing and Implementing Scientific Calculator Tasks and Activities

Barry KISSANE

Although officially approved for use in the Singapore secondary curriculum, including examinations, scientific calculators are still often misunderstood as devices solely for arithmetic computation and even at times regarded as unhelpful for students learning mathematics. Yet the development of calculators in recent decades has been focused on their use as educational devices, and the design of calculators has been heavily influenced by the needs of secondary school students. In this paper, we first consider the educational potential of calculators in education, drawing on a model developed for this purpose. We then focus on the educational design of tasks and activities using calculators for various educational purposes, including the development of mathematical concepts and important processes such as reasoning and modelling. Attention focuses on the design of experiences that incorporate modern calculator capabilities and their affordances for important learning goals, as well as the implementation of these in classrooms. Examples and analysis of some appropriate tasks are provided.

1 Introduction

Although they are almost universally available in schools, especially in environments in which their use in examinations is sanctioned, limited attention seems to have been devoted to considering suitable tasks for scientific calculator use by students and in particular their potential as learning devices. In this chapter, we first highlight the continuing importance of calculators for school mathematics and then briefly introduce a model for their educational use. Drawing on both the relevant literature, as well as extensive developmental work on resources for teachers, a range of calculator tasks is described, and some brief examples of these are offered. The principle purpose of the chapter is to encourage teachers to make better use of a resource that is widely available, but which seems to be frequently neglected in designing classroom tasks.

2 Using Calculators in Education

2.1 *Early experiences*

Calculators first appeared in schools in some countries not long after they were first invented, but only once they became affordable on a wide scale. The precise details of this vary from country to country. In the case of Australia, calculators were being used by schools from the mid-1970s. Their use became almost universal in senior years of secondary schools over a short period of time, especially when various official bodies approved their use in high-stakes examinations in both mathematics and the physical sciences. Their use in earlier secondary years followed shortly afterwards, accelerated by the continuing decline in the (real) price of the devices. Similar phenomena have been characteristic of the appearance of calculators in schools elsewhere, including Singapore.

The original appeal of calculators to schools was almost certainly the opportunities they provided for numerical calculation, which had long enjoyed prominence in school curricula. Early school models in the 1970s provided students with the means to undertake complicated arithmetic tasks, which previously required either written procedures or the use of logarithms. Similarly, tasks that once required students to have access to and expertise in using tables of values were facilitated, as school models routinely provided trigonometric functions, logarithmic and exponential functions, roots and powers. Perhaps surprisingly, calculators with such features were described almost universally as 'scientific' calculators, presumably because calculations of these kinds were important in practical scientific contexts, although they were also important in mathematical and other contexts, and were certainly not restricted in importance to the world of science.

In addition, some early models of scientific calculators also offered limited statistical calculations, for both univariate and bivariate data. While calculators allowed statistical calculation for only small sets of data, these were adequate for school purposes (and, indeed, in many other settings, as small samples are frequently involved when data are collected). Such a capability was especially attractive, as it permitted the calculators to be used to deal with real data, rather than being restricted to artificial data with convenient numbers used to facilitate the otherwise tedious computations involved. Indeed, it is perhaps not surprising that scientific calculators seem to have been adopted first and with more enthusiasm in locations in which statistics and practical measurement were regarded as part of the school mathematics curriculum, and later (and with more reluctance) in settings where this was not the case (such as in Japan or in the state of New South Wales in Australia), where more formal mathematics and Euclidean geometry were more prominent.

2.2 *Curriculum impacts*

As calculators have become more prominent in schools, some curriculum adjustments have become necessary, most obviously in areas related to computation, and most clearly in the mathematics curriculum (less evidently in the science curriculum). Thus, in Australia, for example, successive school mathematics curricula have recognized that arithmetic can be undertaken by students in three different ways: mentally, with written methods or via technology (in this case, a calculator). If all three of these are regarded as important, each requires some curriculum and classroom attention, so that curriculum documents have paid explicit attention to that matter and even offered advice about a suitable balance.

Similarly, when students have routine access to a scientific calculator, work involving trigonometry no longer needs to be restricted to the very few angles (such as 30°, 45° and 60°) for which exact trigonometric ratios are known, measurements involving circles no longer need to have radii that are multiples of seven (to accommodate an approximation of 22/7 for π) and the hypotenuse of triangles no longer needs to be a rational number. While we have become accustomed to such changes over the past thirty years or so, there are still curricula in some countries in which calculators are not used and in which such changes have not been made.

These kinds of adjustments are notably concerned with computation, and for some time mathematics educators and others in many countries have worried that permitting students to use calculators would undermine their learning of mathematics and result in a decline of skills. A good deal of research has been conducted on this point, with very consistent results. Ronau et al. (2011) recently summarized a large body of empirical research work and various meta-analyses, observing—not without some evident frustration—that:

> Few areas in mathematics education technology have had such focused attention with such consistent results, yet the issue of whether the use of calculators is a positive addition to the mathematics classroom is still questioned in many areas of the mathematics community, as evidenced by continually repeated studies of the same topic. As a result, we concluded that future practitioner questions about calculator use for mathematics teaching and learning should advance from questions of whether or not they are effective to questions of what effective practices with calculators entail. (p. 2)

The use of a calculator for computation is now often recognized as part of the curriculum, especially in the area of Number. Thus, for example, The O-level curriculum in Singapore includes the explicit content item in Secondary One, "calculations with calculator" (Ministry of Education, 2012, p. 34). Some curriculum documents are more circumspect, however, and explicitly recognize the importance of not using calculators sometimes (although there is no doubt in other cases that this is implicitly assumed). Thus, many of the Australian Curriculum: Mathematics content statements (Australian Curriculum Assessment and Reporting Authority, 2017) include the phrase "with and without digital technologies". A similar apparent ambivalence persists in other settings as well, even when calculators are accepted as tools for students. No doubt to appease those who persist with the view that calculators might undermine learning (despite research such as that summarized by Ronau et al. (2011) above), the Australian standardized Numeracy tests conducted annually for Year 7 and Year 9 require students to complete two separate test forms, one with a calculator and one without a calculator. (National Assessment Program Literacy and Numeracy, 2017). Kissane (2015) explained that such positions give a clear impression that calculators continue to be regarded mostly as devices for computation.

Other possible uses of scientific calculators for school mathematics are sometimes effectively limited by official prohibitions of calculator capabilities. A common prohibition for many years is that calculators ought not be programmable—although official justifications for this view are rarely offered; in recent years, a growing interest in students learning some aspects of coding in school mathematics has been thwarted to an extent by prohibitions of this kind. Similarly, authorities often prohibit certain calculator capabilities in examinations, and thus effectively ensure that they are not used by students in school. For example, The Singapore Examinations and Assessment Board (2016) prohibits the use of graphics calculators for O-level students and further limits calculators to not have 'programmed' features such as numerical integration and differentiation. While such cautious prohibitions are understandable (at one level), they may have the undesirable consequence of perpetuating an unhelpful view that the main role for a calculator is arithmetic computation, and of limiting access to other more powerful ways in which calculators might contribute to mathematics education.

2.3 *A model for learning*

The earliest calculators were programmed to undertake arithmetical computations, but of course other operations can also be programmed into a calculator—and have been in recent years, as calculators have been developed further to suit the curriculum in schools. So it seems unfortunate that a device that has become widely available continues to be restricted to being used solely for numerical computation. While a concern that students do not lean too heavily on technology to undertake mathematical tasks is appropriate, the other side of such an argument is that it seems inappropriate for students to spend large amounts of their learning time developing by-hand competence which is not quite as efficient or as fast as a cheap calculator. The use of a widely available technology to support the learning of mathematics itself seems a more rational use of this resource as argued by Kissane (2015).

As part of the process of developing support materials (e.g., Kissane & Kemp, 2013) to use calculators for learning mathematics, Kissane and Kemp (2014) developed a four-part model for their educational use that extends beyond mere computation. Details and illustrations of the model are available elsewhere (e.g., Kissane, 2017a), while the model has been used as a framework to consider the educational potential for scientific calculators in Southeast Asia (Kissane, 2016a) as well as for graphics calculators in Junior College (Kissane, 2017b). As the model will be used as a framework for the present paper, a brief description of the four elements in relation to scientific calculators follows.

Representation: Scientific calculators allow mathematical concepts to be represented on a screen. In recent years, calculator representations have improved to become more akin to natural mathematical displays, including the use of conventional mathematical notation. Important learning can be stimulated by representing mathematical ideas in different ways, so that the capacity of a calculator to 're-present' concepts has potential importance for learning. Various representations of fractions and decimals provide a good example of this aspect.

Computation: As noted above, scientific calculators are often regarded entirely as devices for undertaking computation. While this is not the only potential benefit of calculators, it is nonetheless a benefit for some mathematical activity. Thus, practical tasks using everyday measurements can be handled with ease, so that mathematics might move out of the textbook and into the real world of students. For example, this is especially evident for the study of statistics, which can change focus towards students collecting and using their own data. In addition, reducing a need for extensive written computation might create space and time for students to devote to thinking about the mathematics involved in tasks.

Exploration: Scientific calculators allow students to explore some aspects of mathematics by themselves, acting as responsive devices that are controlled by students. In some cases, this component allows teachers to regard the calculator as part of an initial learning process, instead of regarding it as a device to be used after all the necessary mathematics has been learned. Calculators can be used to stimulate and to support open-ended investigations and are helpful devices for students to engage in mathematical experimentation, by themselves or with other students. For example, students can study powers of 2, including non-integral powers, in order to develop a sense of the idea of logarithms to base 2.

Affirmation: The fourth aspect of the model concerns ways in which student thinking can be tested and affirmed—or, indeed, contradicted—through the agency of a scientific calculator. Trivially, a calculator might be used by a student to check that their written answer to a numerical question is correct, providing some reassurance. However, this aspect is intended to reflect a more powerful activity, in which students engage in conjectural thinking, followed by use of the calculator to explore possibilities. For example, in early studies of trigonometry, students might predict—and then use their calculators to check—how the values of the sine function change as an angle increases in size.

A key assumption for this chapter is that productive educational use of scientific calculators by students requires conscious attention to these components of the model; furthermore, it is suggested that this is unlikely to happen naturally for many students without some attention by the mathematics teacher to designing and engineering suitable tasks in the classroom.

3 Task Design

There has been considerable attention by researchers to the general notion of task design in mathematics education in recent years, focused in the recent ICMI Study 22. The associated study conference provided a convenient general description of the idea of a 'task':

> … a task is anything that a teacher uses to demonstrate mathematics, to pursue interactively with students, or to ask students to do something. Tasks can also be anything that students decide to do for themselves in a particular situation. Tasks, therefore, are the mediating tools for teaching and learning mathematics and the central issues are how tasks relate to learning and how tasks are used pedagogically. (Margolinas, 2013, p. 10).

Yet, the particular case of the scientific calculator as an object of attention is surprisingly absent in this work, including in the published ICMI Study (Watson & Ohtani, 2015). Even in the chapter of the study devoted to the use of tools (Leung & Bolite-Frant, 2015), systematic reference to the place of scientific calculators is absent. The recent conceptual and empirical work reported that is associated with calculators has highlighted the significance of more sophisticated technologies such as graphics calculators and CAS calculators, but apparently neglected the less significant technology of the scientific calculator, even though it is a much more prominent device in many countries for students in the early years of secondary school.

Similarly, there appears to be limited attention to scientific calculator tasks in published text materials, such as textbooks, most likely because of unease by publishers or Ministries of Education to appear to give undue prominence to some particular calculators rather than to others, in view of the reality that calculators are commercial products.

Important concepts developed by various researchers, and described well by Trouche and Drijvers (2010) are those of *instrumentation* and the associated process of *instrumental genesis*. These refer to the mechanisms by which an artefact, such as a calculator, becomes an instrument for learning by students, requiring the development of a relationship between the student and the calculator. Tasks are critically involved both in the development of such a relationship and in the ways in which students make use of the calculator as a tool to support their learning, so that attention to what kinds of tasks might be helpful and of how to make use of tasks are central issues to a sound pedagogy.

4 Calculator Tasks

There would seem to be various ways in which calculators might be used by students. In this section, these are briefly described, in a roughly increasing order of significance of the calculator.

Informal. Because calculators are small and generally available, students often choose to use them even though advice to do so has not been provided by their teacher. Although it might be argued that everyday use of calculators in this way by students does not constitute a 'task', it might also be argued that whenever students make use of a calculator, they are engaged in a task of some kind, even if it has not been carefully developed by a teacher. Experience would suggest that the most likely tasks undertaken by students without guidance involve computation of answers to numerical questions, but they might also be engaged in checking answers to questions or even in more significant tasks such as analyzing some statistical data or seeking a solution to an equation by a process of 'guess and check', especially if they were already familiar and experienced with suitable ways of using the calculator for such purposes.

Competence. Some tasks might be developed specifically to support students to use their calculators efficiently. While this is especially the case for more complicated tasks (such as finding and using a line of best fit for bivariate data), it might also be the case for less sophisticated, but important tasks such as setting the calculator to use two decimal places for money computations or learning the appropriate calculator syntax for generating a random integer on an interval. While such tasks are necessary, they do not constitute learning mathematics, and so consume valuable class time. In recent years, calculators have been developed to be more user-friendly, reducing the need for excessive time to be spent on developing calculator competence.

Worksheets. Although worksheets have a variety of meanings, the term is being used here to refer to written tasks for students. Although there are various styles for such tasks, it is not uncommon for a worksheet to be printed on a single sheet of paper and to comprise a number of separate, but possibly related, tasks that involve calculator use in some way. Put simply, a worksheet is a 'sheet' on which some 'work' is provided for students. Worksheets can be developed by teachers, or might in some cases be developed by textbook authors as a supplement or even by calculator manufacturers to support the sound use of the devices. In many cases, tasks on worksheets can be quite short, intended to focus student attention for a few minutes.

Activities. The term 'activity' is used here to refer to an extended task, somewhat longer than typical worksheet tasks, and possibly occupying students for the greater part of a class lesson. In some cases, it might comprise a problem of some kind, for which calculator use is either necessary or especially helpful, while in other cases it might take the form of a piece of mathematical modelling, for which the calculator is an essential resource. Activities are not necessarily printed for distribution to students, but might be introduced as part of a standard lesson or a written version might be projected for all to see. As for worksheets, activities can be developed by teachers for their own class or developed by others for teachers to use.

Replacement units. Some tasks involving calculators might be even more extensive than a single lesson, but might involve a series of lessons in which the calculator is an intrinsic part, perhaps even replacing the 'normal' approach to a topic. For example, students might be engaged in a series of tasks involving the solution of systems of equations, on the assumption that a suitable calculator is routinely available to them. The approach taken might involve regular use of the calculator in various ways, so that the standard approach is effectively replaced. Usually, such work depends on more sophisticated calculators than scientific calculators, such as graphics calculators.

5 Affordances and Design

In this section, attention is drawn to two critical features of calculator activities. Firstly, they should depend on the fresh capabilities afforded by the calculator for students to learn mathematics differently from existing approaches. In the second place, it is important that sufficient teacher guidance be provided, when tasks are intended for the use of others.

5.1 *Affordances*

Computation with calculators is usually well understood; but understanding what calculators afford to students' learning requires a deeper knowledge of both the calculator and mathematics. So the first task for a designer, who may be a teacher, is to understand thoroughly what a particular calculator does, how it does a particular thing and what limitations there are to its capabilities, if any. The concomitant task is to understand how the calculator might then be adapted to help students learn mathematics. To understand thoroughly what a calculator does will require some effort by a designer, or might be assisted by materials developed for that purpose, such as Kissane and Kemp (2013). In general, calculator manuals are unlikely to offer sufficient detail for this purpose.

To give an example, the CASIO fx-82 AU PLUS II calculator, widely used in lower secondary schools in Australia includes a factor command on the keyboard. If this command is executed after an integral result is obtained, the number is replaced by its factors. Figure 1 shows an example of this, with the first screen showing the results of a multiplication and the second screen showing the factored result, which might be regarded as a form of *representation* of the result.

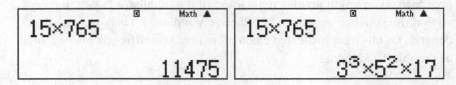

Figure 1. Using a factor command

Such a calculator capability might be used for *computation* purposes, if students needed to find the factors of integers for some reason (e.g. to find greatest common divisors or lowest common multiples), especially if the integers were quite large. However, a more powerful affordance in this case is the calculator's capability to represent products as powers, using the calculator's commands for evaluating indices, as illustrated in Figure 2.

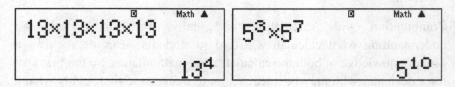

Figure 2. Using a factor command with powers

Figure 2 illustrates that the calculator permits students to see that the representation of numbers as powers using indices can be facilitated through the use of the factor command. Here the concern is not just with computation, but with understanding the representation of numbers using indices and also with seeing the result of multiplying together two numbers that are powers of the same base. In each of the screens in Figure 2, the factor command has afforded students an opportunity to see results differently. However, it is not likely that students will see this opportunity for themselves or take advantage of it in a process of *exploration* to see for themselves what happens when indices are manipulated, unless a suitable task is designed to take advantage of this.

The calculator can be considered as a device on which students can experiment for themselves with indices to see how they work and to provide a stimulus to consider why they work in that way. Laws of indices can be discovered readily by students, given suitable tasks, opportunities and encouragement. The calculator has limitations, however: it will not find factors of negative numbers, or non-integral numbers, and nor will it find factors of very large numbers (for which scientific notation is required to represent the number). An additional limitation in this case is illustrated in Figure 3, which demonstrates that the calculator is restricted to *prime* factors.

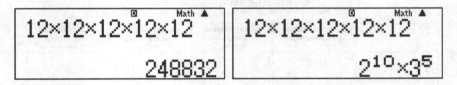

Figure 3. A limitation of prime factors?

Figure 3 illustrates the fourth element of the model, that the calculator allows students to seek *affirmation* of their expectations, as a check on their learning. In this case, a student might be lead to expect that the product $12 \times 12 \times 12 \times 12 \times 12$ will be represented as 12^5, but the calculator's (prime) factorization results in a (surprising) contradiction instead. Rather than regarding this as a problem, we ought perhaps see it as an opportunity for students to learn more about indices, in this case that $(2^2 \times 3)^5 = 2^{2 \times 5} \times 3^5 = 2^{10} \times 3^5$.

This illustration suggests that the educational potential of the calculator relies on a thorough understanding of its capabilities as well as an appreciation of how these are related to important mathematical ideas —that is, its affordances for learning particular aspects of mathematics.

5.2 *Using the model*

The preceding example has served the additional purpose of illustrating how the four elements of the model (Kissane & Kemp 2014) might be involved in thinking about suitable tasks for students to undertake with their calculators. The model can be used to interrogate the relationships between the mathematics that students are learning and the opportunities provided by the calculator to enhance or support that learning.

While the model can be used as demonstrated in the previous section for thinking about a particular topic, a useful strategy for planning tasks is to consider the relationships for a range of topics, through the agency of a planning document such as the calculator affordance table shown in Figure 4.

Calculator affordance table
Enter brief notes to indicate opportunities and examples for calculator use

	Representation ... off numbers, concepts, ...	**Computation** ... of calculations, procedures, ...	**Exploration** ... of relationships, concepts	**Affirmation** ... of predictions, relationships
Number and Algebra				
Geometry and Measurement				
Statistics and Probability				

Figure 4. Calculator affordance table

A table of this kind is one way that a teacher might anticipate various ways in which the calculator can contribute to the learning process over a period of time. In practice, summarizing affordances in this way is a task that can well be undertaken by several teachers together, as affordances recognized by some might not be seen by others.

5.3 *Teacher guide*

To maximize educational value, calculator tasks developed for use in classrooms should be implemented carefully, so that it seems necessary for teachers using them to be given adequate guidance. While this is arguably less necessary for tasks that a teacher plans for use in her own classroom only, it seems important for tasks that are shared with other teachers and critical for tasks that are published for use by teachers because teachers who use these tasks may not have the means to seek clarification directly from the task developer. A teacher guide provides an opportunity to clarify explicitly the role of the calculator for learning mathematics.

The nature of a teacher guide depends in part on the task itself, of course. In some cases, 'answers' to tasks will be sufficient. But in many cases, more substantial help is needed to ensure that tasks have the intended benefits to student learning. At the very least, teachers need advice on the extent to which tasks are intended to be conducted by students individually, or in concert with others. While classes differ from each other, as do contexts, advice about the timing of a task as well as intended follow-up discussions and activities are generally helpful as well.

To illustrate these alternatives, the Exercise tasks in Kissane and Kemp (2013) have only answers provided, and some teachers may even be happy to make these available for students, so that they can confirm their success. However, the Activity tasks include some general advice regarding the nature and purpose of a task, together with answers where appropriate. As a further example, the Investigation tasks in Kissane (2016b) are preceded by a substantial rationale and description of suitable classroom organization, as well as detailed advice in relation to individual tasks. The extra attention provided to teacher guides for these tasks is in part to clarify the nature and purpose of the tasks, on the assumption that the exploratory methodologies involved are less familiar to teachers and students, and hence require more assistance to get maximum benefit.

6 Illustrative Examples

With limited space, only a selection of tasks can be provided to illustrate the possibilities. The following examples have been adapted from freely available publications (Kissane & Kemp, 2013; Kissane, 2016b).

6.1 *Exercises*

Exercises are common kinds of tasks in mathematics textbooks generally, and are generally designed to practice or test a particular skill. Generally speaking, it is assumed that students are taught something first, and then are expected to engage in 'exercises' later to consolidate their learning or to determine whether they have learned what was expected. So, in the case of calculator tasks, a principal purpose of exercises is to ensure that students are able to use their calculator effectively for a particular purpose. Exercises are generally low-level tasks and it is not expected that they will engage students for very long; but they are important tasks to ensure that students develop sufficient expertise with their calculators. Figure 5 shows typical examples of calculator exercises, taken from Kissane and Kemp (2013)

Although the tasks in Figure 5 all have significant mathematical contexts, the main purpose of the tasks is to confirm that students can navigate the calculator efficiently; we will use the CASIO fx-96SG PLUS calculator to illustrate responses to these. Exercise 1 requires students to master the use of both the radical sign and indices, while Exercise 2 requires students to know how to indicate that an angle is measured in radians, even though the calculator is set to degrees, as well as how to enter fractions on the calculator. In both of these cases, the operations (square root and cosine respectively) are entered before the operands, which students also need to know. The appropriate calculator screens for these two tasks are shown in Figure 6.

1. The hypotenuse of a right triangle with shorter sides 7 and 11 can be found by calculating $\sqrt{7^2 + 11^2}$. Give this length as a decimal.

2. Find the cosine of $\dfrac{2\pi}{5}$ radians.

3. Convert the decimal number 62 to hexadecimal notation.

4. A receptionist recorded the number of phone calls received each day of a week, as follows:

Day	Mon	Tue	Wed	Thu	Fri
Number	140	125	134	132	1260

(a) Find the mean and the standard deviation of the daily number of calls for that week.
(b) After completing the calculations, the secretary realized that Friday's data had been wrongly recorded; the correct figure was 126 not 1260. Edit the data to correct this error and then find the mean and the standard deviation of the corrected data.

5. Solve the system of equations: $\quad 2a + 7b = 12$
$$3b = 5a + 1$$

Figure 5. Examples of calculator exercises

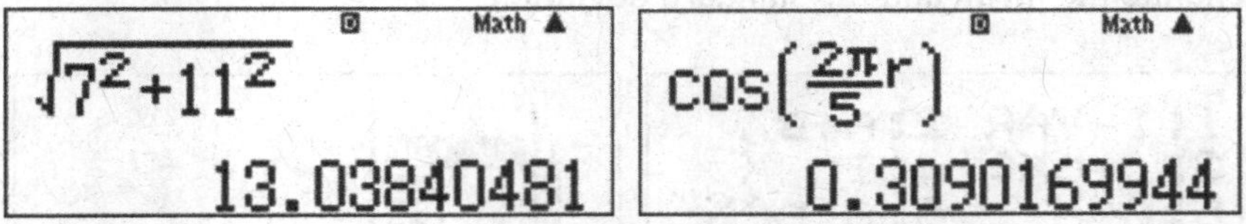

Figure 6. Answers to Exercises 1 and 2

These are essentially computational tasks, facilitated by the calculator's use of natural mathematical displays, so that the input to the calculator is similar to standard mathematical expression. Yet, students still need to find their way around the keyboard commands to evaluate the expressions successfully and efficiently, the main purpose of the tasks. In the case of Exercise 2, for example, students need to know how to indicate that a measure is in radians, rather than the general setting of degrees chosen for regular use.

Exercise 3 requires a little more calculator sophistication, as the students need to change calculator modes to undertake work in bases other than decimal, and then know how to change representations of numbers. These are quick tasks, once learned, but the purpose of the exercise is to ensure that they have been learned, or to provide a stimulus to learn them. Figure 7 shows what is involved, firstly choosing a new calculator mode (BASE-N) and then using calculator commands to obtain the hexadecimal representation 3E of the decimal number 62 (where E represents the decimal number 14).

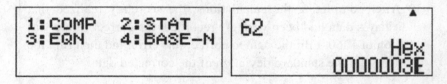

Figure 7. Completing Exercise 3

Exercise 4 also involves students changing modes to Statistics mode, choosing univariate data (described by the calculator as 1-VAR, as shown in Figure 8) to enter the raw data and then using the calculator menus to generate the mean and the standard deviation.

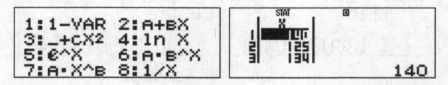

Figure 8. Entering data for Exercise 4

Retrieving the required statistics involves navigating some calculator menus, to obtain the results shown in Figure 9. Although the process is helped by the use of standard representations for mean and standard deviation, students still need to navigate through some calculator menus to obtain the statistics sought.

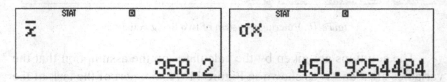

Figure 9. Retrieving statistics for Exercise 4

Similarly the second part of Exercise 4 requires that students retrieve the original data, correct the error described and then repeat the previous steps to find the correct mean and standard deviation. Steps like that are all essential for the use of the calculator for statistics, so the exercise is essentially concerned with the calculator steps. Nonetheless, a classroom discussion of the results shown in Figure 9, especially the large value of the mean in relation to the typical phone call traffic, might suggest to students the importance of careful entry and inspection of both data and results, even if the calculator is handling the arithmetic.

Finally, Exercise 5 also shows another context in which navigation to a different calculator mode (concerned with equations) is needed, but which also demands that students understand how to interpret the calculator syntax. The screens in Figure 10 show, firstly, that students need to recognize that the task involves solving a system of two linear equations, and choose the appropriate kind of equation (the first choice) from the menu. Then, they need to realise that the calculator assumes that the variables are represented by x and y and requires that coefficients are entered in a particular order, and are represented by a, b and c respectively. The given system in Figure 5 uses a and b as *variables*, rather than as coefficients, which requires students to understand the difference, and act accordingly. The second screen in Figure 10 shows the necessary entries to faithfully reflect the system:

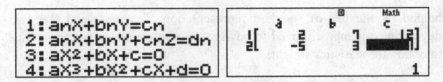

Figure 10. Entering a system of two linear equations

The solutions are given by the calculator on the assumption that the variables are x and y, as shown in Figure 11. Again, part of the task of the student is to recognize that these need to be interpreted as $a = 29/41$ and $b = 62/41$, using the variables given in the original task in Figure 5.

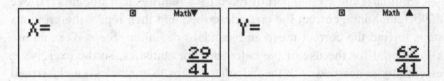

Figure 11. Solutions to the system of two linear equations

While this and the previous exercises are essentially about how to use the calculator, students might incidentally also learn aspects of mathematics; in this case, that the variables chosen to represent a linear system are essentially arbitrary.

6.2 *Worksheets*

The term 'worksheet' has a range of meanings in education, and, indeed sometimes just refers to a sheet of exercises. In this context, I use the term to refer to a single page (a 'sheet') given to students, which includes some substantial mathematical tasks for students, but not restricted to exercises. Good examples are provided in Kissane (2016b), in which the worksheets take the form of sets of related investigative tasks to be completed by students with at least one other student as a partner. For example, Figure 12 shows one of the three investigative tasks from a worksheet concerned with fractions and decimals (adapted here for the CASIO fx-96SG PLUS calculator which is used in Singapore).

In Math mode, when a number is entered into your fx-96SG PLUS, it is usually displayed as a fraction. You can tap the S-D command to represent a number either as a fraction or a decimal.

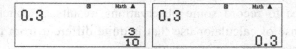

There are many other numbers with the same representation. Here are three more:

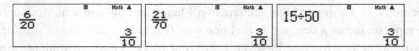

Use your fx-96SG PLUS to find several others. How many can you find? List them below. Discuss this with your partner. Write a brief conclusion:

_____

_____

Figure 12. Investigative worksheet for fractions and decimals

Tasks like this are quite different from the exercises described in the previous section. The intention here is to support younger students learning aspects of decimals and fractions, not just to undertake computations.

Some key elements of this task, described in detail in the introduction to the set of tasks in Kissane (2016b), are that it is self-contained, that it has more open-ness than a closed task like an exercise, and that it assumes that students are working with a partner.

The task is self-contained in the sense that students are given sufficient information to use their calculator, in this case especially involving the use of the S-D ('Standard to decimals') key, which toggles between decimals and fractions. While some particular tasks are suggested, students are encouraged to generate further examples for themselves, so that the calculator is being used as an environment for exploration of the ideas involved, rather than merely as a device to produce a numerical answer. The use of open questions, the reference to a 'partner' and the space to record some observations, results or conclusions all provide a sense of calculator use that is quite different from the closed, individual tasks in the previous section.

This worksheet task exploits the capacity of the calculator to represent numbers in different ways, specifically as fractions and as decimals. It is intended that students will learn that the same number can be represented as a decimal in just one way (excluding the case of trailing zeroes), but as a fraction in several different ways. Although students are frequently taught the skill of 'cancelling' to reduce a fraction to lowest terms, a task of this kind is intended to help them to appreciate what is the same about all the fractions that represent a particular number, and to give a stronger base for the process of cancelling. It does not attempt to provide a logical argument for cancelling, which should of course also be part of their learning at some stage.

Careful analysis of such a task will reveal that it demonstrates all four model elements. While it clearly involves computational tasks, it exploits the capacity of the calculator to represent numbers faithfully, and indeed to re-present them in different forms. The task is exploratory, with expectations for students and their partners to generate further examples for themselves to understand what is happening, and such work will inevitably involve conjectural thinking to affirm or contradict their growing sense of the relationship between fractions and decimals.

6.3 *Activities*

A key distinction between an activity and a task on a worksheet is that it is intended to engage students in an aspect of mathematics for an extended period of time, during which the calculator is a key supporting tool. Thus, where a typical worksheet task might engage students for a few minutes, perhaps to then be the subject of a class discussion, a typical activity might occupy a student or a class for a whole school period, or even longer. Figure 13 shows an example, taken from Kissane and Kemp (2013, p.28):

Many fractions result in recurring decimals, which often show interesting patterns. For example, study the fractions with denominators of 7. Use the calculator to obtain 1/7, 2/7 and then 3/7 and then predict the decimals for 4/7, 5/7 and 6/7, checking your predictions with your calculator.

Examine other fractions with denominators that are prime numbers such as 11ths, 13ths, 17ths and 19ths. Use the calculator to help you to get information in order to look for, and use, patterns.

Figure 13. Recurring decimals activity

As the (brief) teacher guide for this activity notes, recurring patterns are relatively easy to infer from scientific calculator displays for some denominators (such as 3, 7 and 11), a little harder for denominators of 13 (since more than one pattern is involved) and very hard to see for 17ths and 19ths, because the recurring part exceeds the widths of the calculator display, as shown in Figure 14, and thus is not immediately clear.

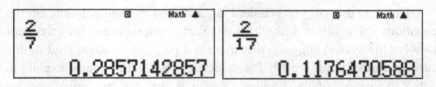

Figure 14. Exploring recurring decimals

Nonetheless, productive and engaging work can be started on a calculator such as the standard calculator used in Singapore to engage with some of the interesting mathematics associated with recurring decimals.

The educational nature of activities depends in part on the actual calculator used, as well as the task assigned. In this case, the use of a different calculator, the CASIO fx-991ID PLUS, used in Indonesia, gives direct access to recurring decimals, representing them on the screen in conventional ways (or, in the case of decimals with more digits, like 17ths, on successive screens), as Figure 15 illustrates.

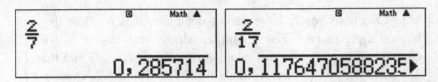

Figure 15. Exploring recurring decimals on a CASIO fx-991ID PLUS

The capacity of a calculator to handle computations quickly leads to many opportunities for productive educational activity. As a second example, consider the activity (Kissane & Kemp, 2013, p. 51) in Figure 16, concerned with the solution of linear systems:

Explain why there is no single solution to the linear system: {3x − y = 7 and 6x − 14 = 2y}. Can you find some other linear systems for which there is no single solution? Check with your calculator.

Figure 16. Activity on solving linear systems

Routine use of the standard algorithm for hand-solution of the equations will result in difficulties. Similarly, routine use of the calculator to solve the system suggests that there is a problem of some kind in this case, as shown in Figure 17. The calculator does not of course explain in detail the nature of the problem, which is that the two equations are dependent on each other, although the result of 'Infinite solutions' gives a strong hint. Rather, the calculator provides an opportunity for students to generate further examples for themselves to look for the patters involved and begin to understand the source of the difficulties. In this case, the arrangement of the equations into the required order for the calculators helps reveal the pattern.

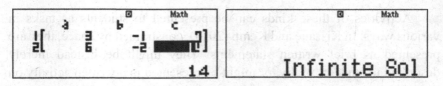

Figure 17. Solving a system of dependent equations

This work might easily be extended to other problematic 2 × 2 linear systems, such as inconsistent equations, where again the calculator provides a helpful tool for experimentation, conjecturing and discovery of underlying patterns.

As a third example, students can use the computational powers of their calculator to explore what happens when data are transformed in various ways. Figure 18 shows one way of initiating such an activity (Kissane & Kemp, 2013, p. 120):

Use your calculator to explore the effects of transforming values. Enter a small data set such as {2, 3, 5, 7, 8} into the calculator and record the mean and standard deviation on paper.
(a)   Add 3 to each data point and find the mean and standard deviation again. Compare with the original statistics.
(b)   Multiply each data point by 4 and find the mean and standard deviation again. Compare with the original statistics.

Figure 18. Studying data transformations

Understanding the effects of data transformations is key to understanding the concepts of mean and standard deviation, as well as being essential to understanding data analysis in later studies (such as transforming normal distributions into standard normal distributions). The calculator allows students to explore the meanings and the consequences of transformations through their own activity, and again will likely involve all four aspects of the model described earlier.

Activities of these kinds can be presented to students as tasks in various ways. In Kissane and Kemp (2013), constrained by space, they are presented as brief written statements. They might be instead merely described to students orally, or might be presented in a written activity on a worksheet with further suggestions and guidance to get students started and to organize their explorations to some extent. Decisions on such matters are best left to the teacher, in light of their understanding of the sophistication of the students and the need for structured support. The key idea, however, is not the form of the activity, but the use of the calculator as a valuable tool to support the students' learning about mathematical ideas through their own experimentation and thinking, as well as discussion with others, rather than in more passive ways.

7 Conclusion

Opportunities for students to learn mathematics through the use of scientific calculators are too frequently neglected, and it is hoped that this chapter will help teachers and others to consider them more carefully. As an always-available device, calculators offer many opportunities to support students learning mathematics. Starting with calculator affordances, and then using a model that permits analysis of student tasks to be undertaken, teachers and others can design a range of worthwhile tasks for students.

References

Australian Curriculum Assessment and Reporting Authority (ACARA) (2017) *Australian Curriculum: Mathematics F-10*. Retrieved 17 August 2017 from https://www.australiancurriculum.edu.au/f-10-curriculum/mathematics/

Kissane, B. (2015) Calculators: Learning, not leaning. *The Australian Mathematics Teacher*, 71(*2*), 34-35.

Kissane, B. (2016a) The scientific calculator and school mathematics. *Southeast Asian Mathematics Education Journal*, 6(*1*), 33-55.

Kissane, B. (2016b) *Investigating mathematics with ClassWiz*. Retrieved on 17 August from http://edu.casio.com/education/activity/ CASIO: Tokyo, Japan.

Kissane, B. (2017a) Learning with calculators: Doing more with less. *The Australian Mathematics Teacher*, 73(*1*), 3-11.

Kissane, B. (2017b) Empowering Junior College students through the educational use of graphics calculators. In B. Kaur & N. H. Lee (Eds.) *Empowering Mathematics Learning*. (pp 51-74). Singapore: World Scientific.

Kissane, B., & Kemp, M. (2013) *Learning Mathematics with ES Plus Series Scientific Calculator*. Retrieved on 17 August from http://edu.casio.com/education/activity/ CASIO: Tokyo, Japan.

Kissane, B., & Kemp, M. (2014) A model for the educational role of calculators. In W.-C. Yang, M. Majewski, T. de Alwis & W. Wahyudi (Eds.) *Innovation and Technology for Mathematics Education: Proceedings of 19th Asian Technology Conference in Mathematics*. (pp 211-220). Yogyakarta, Indonesia: ATCM. [Available for download from http://researchrepository.murdoch.edu.au/24816/]

Leung, A., & Bolite-Frant, J. (2015) Designing mathematics tasks: The role of tools. In Watson, A. & Ohtani, M. (Eds.) *Task Design in Mathematics Education*. (pp 191-225) Switzerland: Springer.

Margolinas, C. (Ed.) (2013) *Task Design in Mathematics Education: Proceedings of ICMI Study 22*. Oxford, UK.

Ministry of Education of Singapore. (2012). *Mathematics syllabus: Secondary one to four*. Singapore: Curriculum Planning and Development Division.

National Assessment Program Literacy and Numeracy (NAPLAN) (2017) *Numeracy*. Retrieved on 19 August 2017 from https://www.nap.edu.au/naplan/numeracy Australian Curriculum Assessment and Reporting Authority.

Ronau, R., Rakes, C., Bush, S., Driskell, S., Niess, M., & Pugalee, D. (2011). NCTM Research Brief: *Using calculators for learning and teaching mathematics*. Retrieved 29 July 2012 from http://www.nctm.org/news/content.aspx?id=31192

Singapore Examinations and Assessment Board (2016). *Guidelines on the Use of Electronic Calculators in National Examinations*. Retrieved 18 August 2017 from http://www.seab.gov.sg

Trouche, L., & Drijvers, P. (2010) handheld technology for mathematics education: flashback into the future. *ZDM Mathematics Education*, *42*, 667-681.

Watson, A., & Ohtani, M. (Eds.) *Task Design in Mathematics Education*, ICMI Study 22, DOI 10.1007/978-3-319-09629-2.

Chapter 7

Engaging the Hearts of Mathematics Learners

Joseph B. W. YEO

In this chapter, I will first propose a theoretical framework on how to engage mathematics students in their hearts. The LOVE Mathematics framework (Linking Opportunities in a Variety of Experiences to the learning of Mathematics) consists of three principles: variety, opportunity and linkage. These principles will address a few common problems in making mathematics lessons interesting, such as, not everyone will find the same thing fascinating; it is not possible to make every part of a lesson engaging; and the enjoyment does not necessarily translate to learning. Then I will describe some resources that teachers can use to make mathematics lessons more exciting. Examples of such resources include the use of amusing maths videos, catchy maths songs, witty maths comics, and intriguing maths puzzles and games.

1 Introduction

Figure 1 shows the Mathematics Framework for the Singapore school mathematics curriculum since 1990 (MOE, 2012). The central goal of the framework is mathematical problem solving. There are five components: concepts, skills, processes, metacognition and attitudes. The focus of this chapter is on attitudes, and in particular, the categories of interest and appreciation to engage the hearts of mathematics learners.

There are some literature on affective education but these are mainly confined to affective variables in general, such as improving students' personal development and self-esteem, interpersonal relationships and social skills, and their feelings about themselves as learners and about their academic subjects (Lang, Katz, & Menezes, 1998). There are also studies on the effect of the students' attitude on test performance (McLeod, 1992), but when it comes to making mathematics lessons interesting and helping students to appreciate mathematics, there is not much literature on this.

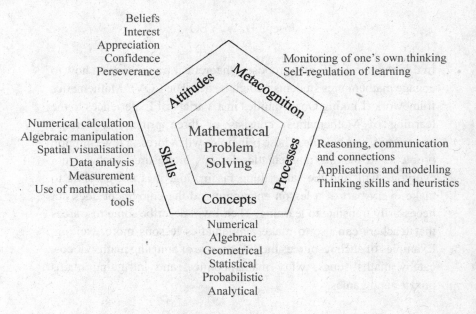

Figure 1. Mathematics Framework

In this chapter, I will develop a theoretical framework on three principles of engaging the hearts of mathematics learners, and then I will describe some exemplars of resources that can be used to make mathematics lessons interesting.

2 Theoretical Framework: the LOVE Mathematics Framework

McLeod (1992) observed that teachers often talked about their pupils' enthusiasm or hostility towards mathematics and that students were just as likely to comment about their feelings towards the subject. These informal observations supported the view that affect played a significant role in mathematics learning and instruction. Reyes (1980) also emphasised the importance of research on student attitudes towards mathematics as such information would enable teachers to teach more effectively and successfully. A good attitude is very important because many research studies have suggested a positive correlation between attitude and achievement in mathematics learning (McLeod, 1992).

The study of attitudes is very complicated partly because there is no common agreement on the definitions of terms. Aiken (1972) used the term attitude to mean "approximately the same thing as *enjoyment*, *interest*, and to some extent, *level of anxiety*" (p. 229). Haladyna, Shaughnessy, and Shaughnessy (1983) used the word attitude as "a general emotional disposition toward the school subject of mathematics" (p. 20). Hart (1989) used the word attitude towards an object as a general term to refer to emotional (or affective) reactions to the object, behaviour towards the object and beliefs about the object. This is similar to Rajecki's definition in 1982 (cited in Hart, 1989). Oppenheim (1992) also subscribed to this view. He believed that attitudes are reinforced by beliefs (the cognitive component) and often attract strong feelings (the emotional component) which may lead to particular behavioural intents (the action tendency component). But Simon suggested the use of affect as a more general term in 1982 (cited in McLeod, 1992). McLeod (1989) expanded this view of affect by dividing the affective variables into the general rubric of beliefs, attitudes and emotions, and he (1992) tried to fit related concepts from the affective domain, such as confidence, anxiety and motivation, into this rubric.

Another reason why the study of the affective domain is complicated is because affective constructs are more difficult to describe and measure than cognition (McLeod, 1992). Gardner observed that past researchers had often avoided studying affective issues because of their desire to avoid complexity (cited in McLeod, 1992). As a result, the impact of Bloom's taxonomy for the affective domain on education (Krathwohl, Bloom, &

Masia, 1964) was very minimal compared to that of his taxonomy for the cognitive domain (McLeod, 1992).

The traditional paradigm for research on affect is very practical as it relies on questionnaires and quantitative methods (McLeod, 1992). Most of these studies focused on students' existing attitude and their effect on other variables such as test performance. There are very few intervention studies, such as on how to change students' attitude.

Therefore, I will propose a theoretical framework on how to engage students in their hearts. There are three principles. The first principle is called *variety*. A usual problem of engaging students in the class is that not all the students will find the same resource interesting. This is understandable because people have different tastes. How to resolve this issue is to use a variety of resources so that hopefully, all the students will find something interesting, although this 'something' may be different for different students. Examples of such resources to engage mathematics learners include amusing maths videos, catchy maths songs, witty comics, intriguing maths puzzles and games, fascinating magic tricks, inspiring stories of famous mathematicians, the celebration of important maths days and years, and interesting real-life examples and applications of mathematics.

The second principle of the framework is called *opportunity*. A usual problem of engaging students in the class is that it is not possible to make every lesson, or most parts of a lesson, interesting, especially when students need to practice procedural skills. But let us look at some of the written comments from my student teachers on my teaching in the SFT (Student Feedback on Teaching) below:

- "Every lesson is engaging and interesting." (2013)
- "Mr Yeo's lessons are always interesting and engaging." (2013)
- "He makes every lesson interesting …" (2014)
- "Makes all lessons interesting and engaging." (2015)
- "Every lesson is engaging …" (2015)
- "He makes his lessons very engaging …" (2016)

- "Interesting and meaningful. I look forwards to all his lessons." (2017)

Many of the statements were sweeping statements along the line of my lessons being interesting or engaging, e.g. "He makes his lessons very engaging…" (2016). None of them ever wrote that 'most' or 'some' of my lessons were engaging. A few of them even went to the extent of writing that 'every' or 'all' of my lessons were interesting or engaging, e.g. "*Every* [italics mine] lesson is engaging and interesting" (2013) and "Makes *all* [italics mine] lessons interesting and engaging" (2015).

But I myself know that it is not possible for all my lessons to be engaging or interesting. So why did many of the student teachers feel that way? I believe it is because I have done enough in most of the lessons for them to feel that all my lessons were engaging. Yes, it is not possible to make every lesson, or most parts of a lesson, interesting, but there is really no need to, as long as the teacher provides ample or enough opportunities in the classroom for the students to be engaged in their hearts. The ultimate aim is to make at least one part of most mathematics lessons interesting.

The third principle of the framework is called *linkage*. Sometimes, a teacher may engage the students using resources that are just for fun and not relevant to the learning of the subject matter. I am not saying that teachers should not do this. For example, I liked the jokes that my teacher told during Chinese lessons when I was a student although the jokes were irrelevant to learning Chinese; but as a result of enjoying these jokes, I looked forward to his lessons and I actually began to like the subject. But whenever possible, the teacher should link the resources to the learning of the subject.

For example, when I was observing a lesson on Pythagoras' theorem as part of a study of school mathematics curriculum enacted by competent teachers in Singapore secondary schools (Kaur, Tay, Toh, Leong, & Lee, 2018), the teacher showed her class a 10-minute excerpt of a two-episode Korean drama called *Splash Splash Love*. The drama was about a senior high school student called Dan Bi, who was not interested in mathematics and somehow travelled back in time to the Joseon era. In order to survive, she had to depend on her high school mathematics and scientific knowledge. The drama was hilarious and the students had a great laugh at the situational jokes and slapstick humour, which when described would no longer be funny to the reader. Why the teacher chose this drama was because there was an incident when the king and his subjects were unable to solve a problem that required knowledge of Pythagoras' theorem, and in the end, Dan Bi solved it for them. But that's all to it in the drama: there was really not much linkage to the learning or application of the theorem.

However, the teacher continued the storyline in the drama by giving the class three problems to do. The first problem described how the king shot a deer across the river at an angle to the river bank. He wanted his hunting trip to be recorded in history, so he needed Dan Bi to help him work out certain information, such as the distance the arrow travelled, which the students had to calculate using Pythagoras' theorem. The second problem continued the storyline of the palace celebrating the king's victory in hunting the deer and the guards wanting to put up a decorative banner between two poles of unequal lengths. Dan Bi was tasked to help the guards calculate the amount of materials used to make the banner, so she needed to find the minimum length of the banner, which the students had to calculate using the same theorem. The third problem continued with the cooks requiring water from the well to cook up a feast for the celebration. Dan Bi came up with a pulley system to retrieve the water from the well, and one of the information she needed was the distance between the two pulleys in the system, which the students had to calculate. What was even more remarkable was that the three problems were in order of increasing complexity and difficulty.

At the end of the lesson, the teacher gave the class homework from the textbook. The following day, one of the students, who had not been handing in homework punctually, unexpectedly handed in her homework on time. It is worth noting that the homework questions were just typical questions on Pythagoras' theorem, not as interesting as the three problems that were done in class, but somehow, the student was motivated by the

Korean drama and the three problems that she just solved the routine homework problems promptly.

The three principles of the theoretical framework, namely variety, opportunity and linkage, can be synthesised to form the LOVE Mathematics framework: Linking Opportunities in a Variety of Experiences to the learning of Mathematics.

3 Exemplars of Resources

To provide ample or enough opportunities in the classroom for students with different tastes to be engaged in their hearts, there is a need for a variety of resources. I have already discussed several interesting real-life examples and real-world applications of mathematics to help students appreciate the relevance of mathematics in Yeo (2010), and the celebration of some important maths days and years in Yeo (2017). In this section, I will show a few exemplars of amusing maths videos, catchy maths songs, witty maths comics, and intriguing maths puzzles and games. As mentioned earlier, situational jokes and slapstick humour in maths videos are no longer funny when you describe them. Moreover, once you read the description of the videos and then watch them, you may not find them amusing anymore because there is no more element of surprise.

3.1 *Amusing maths videos*

An example of such an amusing maths video is called '25 divided by 5 equals 14' (see Yeo, 2018). A man called Billy told his parents that 75% would go to his parents and the remaining 25% would be divided equally among his four siblings and him, which means each of the five children would receive 5%. But his father told him that he was cheating himself because 25% divided by 5 equals 14%. So Billy went on to show his parents on a small blackboard by long division that 25 divided by 5 equals 5 (see Figure 2a; the convention of the symbols is different from what is used in Singapore). But his father showed Billy that 25 divided by 5 equals 14 by long division as well (see Figure 2b): he started by saying that 5 (divisor) would not go into 2 but go into 5, so he wrote '1' on the right hand side; then he subtracted 5 from 25 to obtain 20; after that, he said that 5 (divisor) would now go into 20 to give 4, and he wrote '4' after '1' to obtain '14'.

    (a) Billy's Correct Proof   (b) The Father's Amazing 'Proof'

Figure 2. 'Proof' of 25 divided by 5 equals 14 by long division

Billy proceeded to prove that 5 multiplied by 5 equals 25 (see Figure 3a), but his mother then proved 14 multiplied by 5 equals 25 (see Figure 3b). As you may have noticed by now, each of Billy's proofs is just a one-step proof, which in a way is just stating a fact that proves nothing.

$$\begin{array}{r} 5 \\ \times\ 5 \\ \hline 25 \end{array} \qquad\qquad \begin{array}{r} 14 \\ \times\ \ 5 \\ \hline 20 \\ +\ \ 5 \\ \hline 25 \end{array}$$

    (a) Billy's Correct Proof   (b) The Mother's Amazing 'Proof'

Figure 3. 'Proof' of 25 divided by 5 equals 14 by multiplication

Billy then intended to prove that $14 + 14 + 14 + 14 + 14$ is not equal to 25. He started to count the five '4's (see Figure 4) as he pointed the chalk at them one by one: "4, 8, 12, 16, 20." But his father took over the chalk and counted the five '1's one by one: "21, 22, 23, 24, 25." In the end, Billy was so frustrated that he threw the chalk, which his father had handed back to him, onto the floor.

$$
\begin{array}{r}
14 \\
14 \\
14 \\
14 \\
+\,14 \\
\hline
\end{array}
$$

Figure 4. 'Proof' of 25 divided by 5 equals 14 by addition

The main goal of letting students watch the video is not for most of them to laugh at the ridiculous 'proofs'. The last principle of the LOVE Mathematics framework proposes that the teacher still needs to link the video to the learning of mathematics. Therefore, the teacher needs to ask the class to explain what are wrong with the three 'proofs'.

The first 'proof' is not easy to explain. So we will start with the second proof. When multiplying 5 times 1 in the number 14, Billy's mother wrote '5' in the ones place (see Figure 3b), when it should be '5' in the tens place (see Figure 5a) because the digit 1 in the number 14 is actually 10. But do students really understand the procedure in Figure 5a, that the '5' in the fourth line actually stands for 50 because it is in the tens place? One way to test for understanding is to add a '0' to the '5' in the fourth line (see Figure 5b). If a student tells the teacher that '50' is wrong because it should be '5', then it will reveal that the student has a serious misconception about the procedure. Of course, if a student says that by convention, we do not put a zero after '5' but the '5' actually represents '50', then we know that the student has understood the procedure correctly.

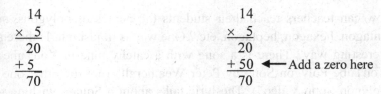

(a) Correct Procedure   (b) Testing for understanding

Figure 5. Correct multiplication procedure

Just like the second 'proof', the third 'proof' has the same issue with place values. Billy's father treated the digit 1 in the number 14 as '1' one, instead of '1' ten. As a result, after Billy has counted to 20 by adding the five '4's (see Figure 4), his father continued to count the five '1's as '1's, rather than '10's.

Similarly, the problem with the first 'proof' is still the issue of place values. When Billy's father said that 5 (divisor) would not go into 2 but go into 5, and he wrote '1' on the right hand side (see Figure 2b), this number '1' actually represents '1' one, or 1 group of 5. After he subtracted 5 from 25 to obtain 20, and he said that 5 (divisor) would now go into 20 to give 4, this number '4' represents '4' ones, or 4 groups of 5. By right, 1 group of 5 plus 4 groups of 5 will give 5 groups of 5, i.e. 25 divided by 5 is still 5. But he wrote the number '4' after the number '1' to give 14. In other words, the number '1', which is supposed to be '1' one, has now become '1' ten.

The main lesson from this amusing video is to teach students the importance of the place value system. If they get the place value wrong, it will end up with three ridiculous 'proofs' that 25 divided by 5 is 14.

For more samples of amusing maths videos, see Yeo (2018). Although my online repository is mainly for secondary school mathematics, some of the resources are actually for primary school students.

3.2 *Catchy maths songs*

How can teachers teach their students the names of polygons, such as pentagon, hexagon, heptagon, etc.? One way is to just tell. Is there a more interesting way? There is a song with a catchy tune in YouTube called "YouTube Polygon Song by Peter Weatherall" (not the other one called "Polygon Song Video"). The lyric talks about a Square wishing to be a Pentagon, a Hexagon, etc., and not 'just a boring square'. In the end, it underwent cosmetic surgery and became a Decagon.

Figure 6 shows an activity in a Singapore textbook (Yeo, Teh, Loh, Chow, et al., 2013) where students will listen to the song and write down the names of the polygons. The teacher needs to pause the video at appropriate junctures for the class to copy the names of the polygons because the names will disappear too fast in the video. In the textbook, heptagon and nonagon are given because they are not in the syllabus: students are required only to learn the names of *n*-sided polygons for *n* = 5, 6, 8 and 10 (of course, students have already learnt the terms 'triangles' and 'quadrilaterals').

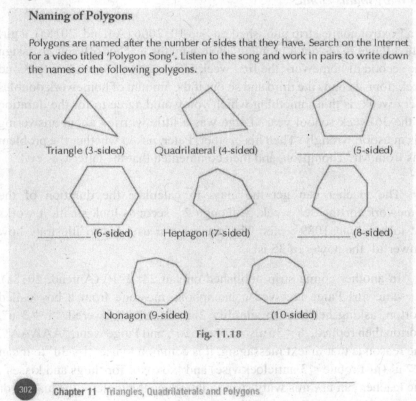

Naming of Polygons

Polygons are named after the number of sides that they have. Search on the Internet for a video titled 'Polygon Song'. Listen to the song and work in pairs to write down the names of the following polygons.

Triangle (3-sided)   Quadrilateral (4-sided)   _____ (5-sided)

_____ (6-sided)   Heptagon (7-sided)   _____ (8-sided)

Nonagon (9-sided)   _____ (10-sided)

Fig. 11.18

302   Chapter 11   Triangles, Quadrilaterals and Polygons

Figure 6. Textbook activity on naming of polygons (reproduced with permission)

One issue with the song is that all the polygons that are shown in the video are regular polygons because of the storyline: the Square would aspire to become a regular polygon rather than an irregular shape. But the textbook activity has included some irregular polygons to make it clear to the students that the names apply to any polygons: whether regular or irregular. Another 'issue' with the song is that the lyric says "my sides equal four", which means that the Square has four sides. But if the students did not listen carefully, they might misunderstand that the lyric is saying that all the four sides are equal. For more samples of catchy maths songs, see Yeo (2018).

3.3 *Witty maths comics*

In a Foxtrot comic strip published on Sep 10, 2006 (Amend, 2018a), a girl called Paige reads out a problem about a maths teacher offering to assign one second of homework the first week of school, two seconds the second week, four seconds the third and so on; if the amount of homework doubles every week, is this something which you would agree to for the duration of the 36-week school year? Paige was a little worried about answering this question wrongly. Then his brother, Peter, asked whether the problem was from Mr. Thompson, and then commented that the latter was evil.

The teacher can get the class to calculate the duration of the homework for the 36th week. Although 2^{35} seconds look small, it works out to be about 1089 years. The teacher can use this to illustrate how 'powerful' the power of 35 is!

In another comic strip published on Jan 24, 2010 (Amend, 2018a), the same girl Paige received a handphone message from a boy called Morton, asking her how to simplify $2i < 6u$. She answered, "$i < 3$ u". Morton then replied, "$i < 3u$ also. xoxoxoxo", and Paige went, "AAAAA!" The reason is that in text messaging, it is common to use, '$i < 3u$' to mean 'I ♡ u' (just rotate < 3 anticlockwise) and 'xoxoxo' for 'hugs and kisses'. The teacher can use this witty strip to introduce the topic of inequality, or after teaching the topic. Sometimes, a comic strip is just for fun. For more samples of witty maths comics, see Amend (2009, 2018a, 2018b).

3.4 *Intriguing maths puzzles and games*

How can teachers make practice questions more interesting? One way is to convert it to a simple puzzle. Figure 7 shows two samples from two Singapore textbooks (Yeo, Teh, Loh & Chow, 2013; Yeo, Teh, Loh, Chow, et al., 2013). The first puzzle has a riddle that students can find the answer to by doing the practice questions. In the second puzzle, students are to draw a path for the monkey to the banana passing through tiles that contain surds.

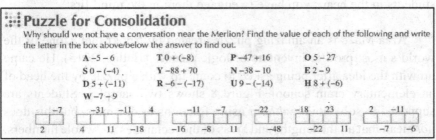

⋮⋮ Puzzle for Consolidation

Why should we not have a conversation near the Merlion? Find the value of each of the following and write the letter in the box above/below the answer to find out.

A $-5-6$	T $0+(-8)$	P $-47+16$	O $5-27$
S $0-(-4)$	Y $-88+70$	N $-38-10$	E $2-9$
D $5+(-11)$	R $-6-(-17)$	U $9-(-14)$	H $8+(-6)$
W $-7-9$			

-7		-31	-11	4		-11	-7		-22		-18	23		2	-11
2	4		11	-18		-16	-8		11	-48		-22	11		-7	-6

A monkey can only walk on tiles containing surds. Help the monkey find the banana by shading its path. (The monkey can only move along a row or a column; it cannot move diagonally across tiles.)

🐵▶	$\sqrt{2}$	$\sqrt{3}$	$\sqrt[3]{6}$	$\sqrt{5}+2$	2019	$\dfrac{\sqrt{2}}{3}$	$\dfrac{3-\sqrt{5}}{2}$
	$\sqrt{5}$	2013	$\sqrt{4}$	$\sqrt{10}$	$1-\sqrt[3]{7}$	$7\sqrt[3]{9}$	$\sqrt{100}$
	$\sqrt[3]{7}$	$1+\sqrt{2}$	2014	$\sqrt{25}$	$\sqrt[3]{8}$	2017	$\sqrt[3]{12}$
	$\sqrt{9}$	$\sqrt{8}$	$\sqrt[3]{27}$	🌙	$\sqrt{27}$	$\sqrt[3]{16}$	$6\sqrt{6}$
	$\sqrt{7}+\sqrt{4}$	$2\sqrt{5}$	$\sqrt{49}$	2015	$\sqrt[3]{64}$	$\sqrt{\dfrac{27}{3}}$	$\sqrt[3]{36}$
	$\dfrac{2+\sqrt{3}}{5}$	2018	$\dfrac{2\times\sqrt{7}}{\sqrt{7}}$	$4\sqrt{8}$	$\dfrac{1}{2}\left(2-\sqrt{3}\right)$	$\dfrac{\sqrt[3]{10}}{3}$	$\sqrt{2016}$
	$\sqrt{11}$	$\sqrt[3]{10}$	$\sqrt[3]{2}-1$	$\sqrt[3]{25}$	$\sqrt{9}+\sqrt[3]{27}$	$\sqrt{81}$	$\sqrt[3]{1416}$

Figure 7. Simple Maths Puzzles (reproduced with permission)

However, some students will not find the above maths puzzles, video, song and comics funny at all. For them, what interest them is still solving some good problems. For the maths video in Section 3.1, I have explained how the teacher can get the students to explain why the three 'proofs' are wrong, and similarly for the first comic strip in Section 3.3. But there may not be anything in the maths song in Section 3.2, or the second comic strip in Section 3.3, or the first two puzzles in this section, that may interest these students. So we will look at some other types of maths puzzles and games that these students may find intriguing to solve. To engage these students in the heart, you have to engage them in the mind first.

Area Maze is an amazing puzzle created by Naoki Inaba, one of the world's most prolific inventors of logic puzzles (Bellos, 2015). He came up with the idea after being asked to come up with a puzzle by the head of an elementary cram school. Figure 8 shows two samples. Students are supposed to solve them *without* using fractions or decimals, but this does *not* mean that all the lengths and areas of the rectangles are whole numbers. The rectangles are usually not drawn to proportion to prevent guessing.

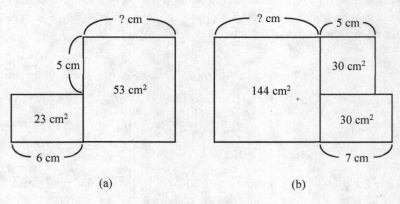

Figure 8. Area Maze

In the first sample (see Figure 8a), students must not divide 23 cm^2 by 6 cm to find the breadth of the left rectangle because the breadth will be a fraction. Instead, students can draw a rectangle (let's call it A) on top of the left rectangle (let's call it B), with length and breadth 6 cm and 5 cm respectively. Then the area of Rectangle A is 30 cm^2. Thus the total area of Rectangles A and B is $23 + 30 = 53$ cm^2, which is equal to the area of the right rectangle (let's call it C). Now, Rectangles A and B form another rectangle (let's call it D). Since both Rectangles C and D have the same area and they share a common side, then they are identical (or congruent). Therefore, the missing length of Rectangle C is 6 cm. I will leave the second sample (Figure 8b) for the reader to solve. For more samples, see Bellos (2015): the last sample looks complex, but the second last sample is actually the most difficult one.

For another intriguing puzzle, the reader can search for KenKen in the Internet, which I do not have the space to describe here.

We will now turn our attention to logical games. Figure 9 shows two logical games called Rush Hour and Tipover respectively. For Rush Hour, the player is required to move the red car out of the rush-hour jam to the little opening on the right of the platform. All the cars and trucks can only move forward and backward, but not make left or right turns. There are 40 challenge cards which give the starting positions of the cars and trucks: 10 at beginner level, 10 at intermediate level, 10 at advanced level and 10 at expert level. The reader may have seen this game online or as an app, but the beginner levels are too difficult while the expert levels are not challenging enough. It is not easy to design the challenges but the original game described here has developed the various challenges appropriately.

The second game in Figure 9, Tipover, requires the player to move the red man from the starting position given by the challenge card to the red crate. The man cannot step onto the floor of the warehouse, but he can only move along the top of the crates as long as the crates are arranged next to each other and not diagonally across. The man can also tip over the crate that he is standing on, provided the crate falls flat onto the warehouse floor and not onto another crate. Then the man can walk along the fallen or horizontal crate. There are 40 challenge cards: 10 at beginner level, 10 at intermediate level, 10 at advanced level and 10 at expert level.

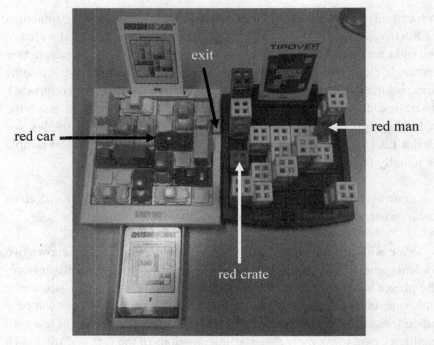

Figure 9. Rush Hour and Tipover

The purpose of letting students play such games is to develop problem solving strategies such as trying to figure out the endgame and then working backwards, examining all the possible moves at each stage and systematically exhaust all the options, and spatial visualisation. Such heuristics and thinking skills are important for mathematical problem solving as well. Thus the main goal of letting students play these games is not just for their enjoyment, but the teacher should make explicit these strategies when the students are playing the games.

I have let secondary school students and trainee teachers play such games before. For students, a good time to let them play is after exams when they have nothing much to occupy them. Some of them were hooked on the games and they refused to go home even after the class had ended. Of course, there will be students who are not interested in these games because people have different tastes.

5 Conclusion

The LOVE Mathematics framework proposes that we should (a) use a variety of experiences to interest different students because not every student will find the same thing intriguing, (b) provide ample opportunities to engage the students, e.g. in at least one part of most lessons; and (c) link the experiences to the learning of mathematics because the main objective is not to make students laugh. This chapter has also described some of these resources that teachers can use to engage their students in the hearts. However, there is a need for more research to study the effectiveness of such strategies.

Acknowledgement

Part of this chapter refers to data from the research project "A study of school mathematics curriculum enacted by competent teachers in Singapore secondary schools" (OER 31/15 BK), funded by the Office of Educational Research, National Institute of Education (NIE), Nanyang Technological University, Singapore, as part of the NIE Education Research Funding Programme. The views expressed in this chapter are the author's and do not necessarily represent the views of NIE.

References

Aiken, L. R. (1972). Research on attitudes toward mathematics. *Arithmetic Teacher, 19*, 229-234.

Amend, B. (2009). *Math, science, and Unix underpants: A themed FoxTrot collection by Bill Amend*. Kansas City, Missouri: Andrews McMeel.

Amend, B. (2018a). *FoxTrot*. Retrieved from http://home.assets.gocomics.com/foxtrot

Amend, B. (2018b). *FoxTrot*. Retrieved from http://www.foxtrot.com

Bellos, A. (2015). *Can you solve it? Are you smarter than a Japanese schoolchild?* Retrieved from http://www.theguardian.com/science/2015/aug/03/alex-belllos-monday-puzzle-question-area-maze-smarter-than-japanese-schoolchild

Haladyna, T., Shaughnessy, J., & Shaughnessy, J. M. (1983). A casual analysis of attitude toward mathematics. *Journal for Research in Mathematics Education, 14*, 19-29.

Hart, L. E. (1989). Describing the affective domain: Saying what we mean. In D. B. McLeod & V. M. Adams (Eds.), *Affect and mathematical problem solving: A new perspective* (pp. 37-45). New York: Springer-Verlag.

Kaur, B., Tay, E. G., Toh, T. L., Leong, Y. H., & Lee, N. H. (2018). A study of school mathematics curriculum enacted by competent teachers in Singapore secondary schools. *Mathematics Education Research Journal, 30*, 103-116.

Krathwohl, D. R., Bloom, B. S., & Masia, B. B. (1964). *Taxonomy of educational objectives: The classification of educational goals. Handbook II. Affective domain.* New York: David McKay.

Lang, P., Katz, Y. J., & Menezes, I. (Eds.). (1998). Affective Education: A Comparative View. London: Cassell.

McLeod, D. B. (1989). Beliefs, attitudes, and emotions: New views of affect in mathematics education. In D. B. McLeod & V. M. Adams (Eds.), *Affect and mathematical problem solving: A new perspective* (pp. 245-258). New York: Springer-Verlag.

McLeod, D. B. (1992). Research on affect in mathematics education: A reconceptualization. In D. A. Grouws (Ed.), *Handbook of research on mathematics teaching and learning* (pp. 575-596). New York: MacMillan.

Ministry of Education of Singapore. (2012). *Mathematics syllabus: Secondary one to four.* Singapore: Curriculum Planning and Development Division.

Oppenheim, A. N. (1992). *Questionnaire, design, interviewing and attitude measurement* (New ed.). London: Continuum.

Reyes, L. H. (1980). Attitudes and mathematics. In M. M. Lindquist (Ed.), *Selected issues in mathematics education* (pp. 161-184). Berkeley, CA: McCutchan.

Yeo, J. B. W. (2010). Why study mathematics? Applications of mathematics in our daily life. In B. Kaur & J. Dindyal (Eds.), *Mathematical modelling and applications* (pp. 151-177). Singapore: World Scientific.

Yeo, J. B. W., Teh, K. S., Loh, C. Y., & Chow, I. (2013). *New Syllabus Additional Mathematics* (9th ed.). Singapore: Shinglee.

Yeo, J. B. W., Teh, K. S., Loh, C. Y., Chow, I., Neo, C. M., & Liew, J. (2013). *New Syllabus Mathematics 1* (7th ed.). Singapore: Shinglee.

Yeo, J. B. W. (2017). Use of open and guided investigative tasks to empower mathematics learners. In B. Kaur & N. H. Lee (Eds.), *Empowering Mathematics Learners* (pp. 219-248). Singapore: World Scientific.

Yeo, J. B. W. (2018). *Maths songs, videos and games for secondary school maths.* Retrieved from http://math.nie.edu.sg/bwjyeo/videos

Chapter 8

Developing Interaction Toward the Goal of the Lesson in a Primary Mathematics Classroom

Keiko HINO

Mathematics instruction needs to be understood from multiple perspectives and involves a variety of components. One of these components is the goal of the lesson(s). The purpose of this chapter is to explore the role of the goal of the lesson that the teacher has in mind so as to construct and manage actual classroom interactions. By examining several studies on classroom interaction, an emerging perspective that places the teacher in the key role to stimulate and enhance students' mathematical thinking in the classroom is demonstrated. Subsequently, by employing a framework of guided focusing pattern (Funahashi & Hino, 2014), classroom interactions in two mathematics lessons taught by an experienced teacher are illustrated in terms of how they are channeled toward the goal. Furthermore, the interaction from the teacher's perspective is explored and how the teacher monitors his teaching actions is examined. The analysis revealed that the teacher had a clear goal of the lesson and moreover, a network of goals that cover the entire unit of study. Such goals, in conjunction with his pedagogical value of interpersonal relations, had a profound influence on his managing a variety of students' responses. He highlighted important mathematics and guided the students to be involved in the process of interpreting and clarifying simple language or non-standard methods, which he claimed is essential to develop students' comprehension.

1 Introduction

The goal or objective of a lesson is a key component that structures mathematics instruction. For example, in the study of lessons, goal-setting is a core activity of teachers. The lesson study cycle (e.g., Lewis & Hurd, 2011, cited in Fujii, 2015) contains *study curriculum and formulate goals* as the first phase of the cycle. In this regard, the goal is long-term and covers a whole year or even a longer period and therefore, is stated in general terms. In the second phase, *plan lessons*, teachers formulate goals of the lesson(s) by connecting them with their long-term goal as well as to the curricular and content goals of mathematics. Teachers' planning of tasks used in the lesson, sequence of activities and method of evaluation are all closely linked to the goal of the lesson. In the phase, *reflect with others*, teachers debrief the lesson from the viewpoint of the goals of the lesson they had set and revise the lesson plan so that students can reach the goal in a more preferable manner. In this phase, teachers may modify or change the goal of the lesson. The quality of the goal is examined repeatedly because it reflects the *educational value* (Fujii, 2015) of teachers.

In this chapter, the focus is on the goal of the lesson that the teacher actually has in mind when conducting lessons. Classroom interaction between the teacher and students is at the core of the discussion. Furthermore, the aim is to explore the roles of lesson goals in the interaction. To achieve this aim, studies that place the teacher as playing an important role so as to stimulate and enhance students' mathematical thinking in the classroom are reviewed. In these studies, different viewpoints of how to shift students' attention to important mathematics are proposed. In a case study of a Japanese teacher that follows, I analyze two consecutive lessons in a fifth grade classroom and illustrate classroom interactions in terms of how they are channeled toward the goal. The questions pursued in the analysis involve (i) what goals the teacher had in mind, and (ii) how those goals functioned in the interaction with the students.

2 Teacher's Role in Classroom Interaction

2.1 *Selected review of research*

It has generally been accepted that high qualities of classroom interaction and discursive practice are essential to foster effective student learning. Nevertheless, identifying effective discursive practices in mathematics, a domain-specific area, is according to Walshaw and Anthony (2008), still in its formative stage. They presented a comprehensive review of what mathematics teachers actually do to promote mathematical discourse that allows students to achieve desirable outcomes. They synthesized the literature around four activities: clarifying discourse participation rights and responsibilities; scaffolding students' ideas to move thinking forward; fine-tuning mathematical thinking through language; and shaping mathematical argumentation. All these activities involve crucial roles of the teacher so as to control classroom interactions. For example, the teacher needs to differentiate between students' ideas purposefully and to utilize their ideas as a springboard for developing new related knowledge. The teacher also needs to bridge students' intuitive understanding with mathematical conventions and to direct them toward modes of thinking that are characterized by precision and brevity. Walshaw and Anthony emphasized, "[R]esearch quite clearly demonstrates that pedagogy that is focused solely on the acceptance of all answers and solutions does not strike at the core of what mathematics discourse truly entails" (2008, p. 539).

The vital role of the teacher is also suggested in a study conducted by Wood, Williams, and McNeal (2006) that explored the relationship between classroom interaction and students' mathematical thinking. They described the importance of particular modes of participation in classroom interaction by highlighting the advantage of interaction in which "children's thinking was extended, pulled together, or strengthened by argument" (p. 248) with the teacher and classmates. Accordingly, the role of the teacher in exerting leadership and intervening in students' reasoning was highlighted. Purposeful support of argumentation and inquiry enable students to ask questions, make decisions, and share the validity and meaning of solutions offered by their peers, which are deemed as higher-level mathematical thinking.

An area of study that addresses the dilemma of the teacher's act of telling and not-telling has offered further insights into teachers' significant roles in classroom interaction. Chazan and Ball (1999) rethought the teacher's act of telling students. They expanded the definition of telling to include teaching actions such as a teacher rephrasing students' comments for the whole class and a teacher employing a new voice through questions and comments. Lobato, Clarke, and Ellis (2005) highlighted the issue of downplaying teaching as telling and reformulated telling as *initiating*. Initiating consists of a set of teaching actions that serve the function of stimulating students' mathematical constructions through the introduction of new mathematical ideas into a classroom conversation. Three defining features of initiating are *intention*, *action*, and *interpretation*. The concept of initiating emphasizes the teacher's intentional prompt of consistency and making sense in the discussion, the teacher's variety of actions for expanding students' ideas and introducing new concepts and representations, and the teacher's constant interpretation of and reflection on students' reactions to his/her actions.

Lobato, Hohensee, and Rhodehamel (2013) investigated how the process of focusing on important mathematics has a decisive role in the development of students' reasoning. By analyzing two classrooms in which the same mathematics content was taught from the perspective of what the lesson focused on and in what way, they argued that if students notice a certain feature in complex mathematical features, it provides the foundation for future products of students' reasoning. On the basis of the results, they asserted, "… teachers can play an important role in directing students' attention toward or (unintentionally) away from what is centrally important for students to notice for a given topic" (p. 845).

Selling (2016) investigated three mathematics classrooms in middle and high schools in the United States in relation to how teachers make mathematical practices explicit in classroom discourse. The focus of this study involved the period of class discussion with respect to how teachers initiate, sustain and reprise the discussions, and in particular, the *reprising* these teachers facilitated in the class discussion. Selling (2016) discovered eight types of teacher actions that made mathematical practices explicit. They included naming the mathematical practice in which students had just engaged and highlighting aspects of student engagement in mathematical practices.

In summary, a teacher's significant roles in classroom interaction and discourse have been indicated; these include scaffolding students' ideas so as allow them to extend and move forward as well as initiating, focusing and highlighting new mathematical ideas and thinking, and important mathematical practices. The researchers have pointed out or suggested that teachers' intentions are an important feature of explaining and interacting with their learners. Walshaw and Anthony (2008) thus described this:

A successful teacher of mathematics will have both the *intention* and the *effect* to assist pupils making sense of mathematical topics. Moreover, the effective teacher is able to make sense of students' conceptual understandings and is able to determine where those understandings might be heading (p. 539, italics original).

2.2 *Guided focusing pattern*

To examine classroom interaction that makes the development of student's mathematical thinking and understanding possible, Funahashi and Hino (2014) explored classroom social interaction, in which students' thinking is placed at the center, along with the teacher's role in balancing students' thinking, with the goals that the teacher wishes to achieve during the lesson. The framework is called a *guided focusing pattern* and is intended to describe the interactive process in which new mathematical content is introduced to students.

The guided focusing pattern comprises four phases. In the first phase, namely, the phase of *proposing the problem (proposing phase)*, the teacher assigns a task for the day. The task often contains open questions. The teacher encourages students to develop their own views of how to arrive at the answer and invites them to share their views and approaches on how to explore the task.

In the second phase *eliciting students' ideas (eliciting phase)*, students present their solutions and the teacher accepts and/or elaborates on them. The teacher asks questions that elicit multiple ideas from students. Furthermore, the teacher emphasizes important points found in students' ideas that can be employed in subsequent phases.

In the third phase *focusing the object of examination (focusing phase)*, the teacher focuses on an important idea that students have proposed during the eliciting phase. By comparing different solutions, the teacher attempts to make the students' ideas the object of examination. Blackboard writing is one of the key prompts. If students do not spontaneously produce the idea expected by teacher, the teacher facilitates the generation of this idea or leads students to it by devising questions.

The fourth phase *formulating the result (formulating phase)* concludes the interaction. The teacher formulates results and/or approaches in general terms and does not limit them to the given problem. Sometimes the teacher asks students to summarize a point from the discussion by using their own words.

Of the four phases, the focusing phase and formulating phase are particularly important because it is in these phases that students come to pay more explicit attention to the important ideas that become the foundation of mathematical content. The analyses of several lesson episodes have revealed the teacher's decisive role in building and refining students' foci, which are closely linked to the objectives of the lesson (Funahashi & Hino, 2014; Hino & Koizumi, 2014).

3 Data Source, Collection, and Analysis

The data analyzed in this chapter have been taken from two consecutive lessons that dealt with comparing fractions in a fifth grade classroom in a university-affiliated primary school in Tokyo. The lesson was conducted as part of the Learner's Perspective Study-Primary (Shimizu, 2011). These two lessons were the first two lessons of the entire 16 recorded lessons on the chapter that deals with addition and subtraction with unlike denominators.

The researchers collected a large amount of data from the lessons as well as from an interview with the teacher and four focus students (Fujii, 2013). The interview was a stimulated-recall type. In other words, interviewees were shown the video of their recorded lesson and asked to make comments on their important learning or teaching in the case of the teacher moments (see also Clarke, 2006). All of the lessons and interviews were transcribed. In the analysis of this chapter, lesson records and teacher interviews as sources of data were utilized.

In the analysis, the teacher's views on the mathematical background of the topic taught and educational goal of the first two lessons were first identified. Secondly, the two lessons were classified according to the framework of guided focusing pattern. Thirdly, focus was given to the four focusing phases and the data were examined in terms of shared focus of attention among the teacher and the students in the whole class discussion. By employing the teacher's interview data, the comments made by the teacher on the important teaching moments were examined as the teacher's thought processes behind his instructional moves to channel the students toward the lesson goal.

4 Role of the Teacher in the Focusing Phases

4.1 *The teacher's goal of the lessons*

The teacher's views on mathematical background of the topic taught. Mr. Taka (all names in this chapter are pseudonyms) was an experienced teacher in primary school. He had been active in lesson studies by proposing his ideas of mathematics teaching and contributing to mathematics education in his community nationwide.

In the teacher interviews, Mr. Taka emphasized the idea repeatedly of finding a common unit fraction when comparing fractions. Such comments were classified into three groups. Firstly, the idea is important so as to deepen the students' understanding of the comparison of fractions because the students often grasp equivalent fractions merely as symbol manipulation by multiplying the same number in upper/lower parts of the fraction. He especially stressed the importance of drawing figural representations to connect symbol manipulation and its meaning on quantity. Secondly, Mr. Taka spoke about the "real meaning" of the calculation of fractions with unlike denominators, that is, we can calculate fractions because the units are the same. He emphasized the fundamental idea of unit that underlies the different mathematical content of calculations. Mr. Taka said, "As for the calculation with whole numbers, the unit is one. As for the calculation with decimal numbers, the unit is 1/10. This time, we apply the same idea to fractions with unlike denominators." Thirdly, Mr. Taka even connected the significance of unit and unit fractions to the defining conditions of numbers in primary mathematics. For something to be accepted as a number, it must satisfy several conditions, namely, it can be compared and added, subtracted, multiplied, and divided. Finding a common unit fraction and rescaling the original fractions by the unit fraction enables the two fractions to be compared and calculated. For the students, it is deemed to be an important discovery that once the rescaling is successful, one can compare the fractions and add or subtract fractions in the same way as previously learned with whole numbers, decimal numbers, or fractions with like denominators.

In this way, the concepts of unit and unit fractions are considered to be connected critically with building students' understanding of fractions as numbers. He consistently focused on these concepts in the lessons.

The teacher's goal of the first two lessons. In the interview, Mr. Taka said that the goal of Lesson 1 (L1) was "to understand that one can compare [fractions] once a common unit is found." When asked about the reason for this goal, he said, "When the students learn addition and subtraction of fractions with unlike denominators, it connects to the idea that one can calculate fractions when unit fractions are the same." With reference to the goal of Lesson 2 (L2), Mr. Taka said, "[it is] to understand that one can compare [fractions] when either denominators or numerators are the same and if it is possible, the students find a way of making either denominators or numerators the same." He added that he wanted the students to understand why it is possible to compare fractions by finding the least common multiple rather than just applying the method.

4.2 *The guided focusing pattern in the two lessons*

In Figure 1, the placement and duration of the four phases of the guided focusing pattern in L1 and L2 are portrayed.

In L1, Mr. Taka presented three different fractions 2/3, 3/4, 2/4, and asked the students to consider which one was bigger. He asked the students to speak about what they noticed at a glance. The light gray band in Figure 1 demonstrates when the students were proposing their observations (eliciting phase). Then Mr. Taka directed the students' attention to explaining why 2/3 > 2/4. The focused discussion lasted about 15 minutes; this is shown by the dark band (focusing phase). Subsequently, Mr. Taka posed a question by pointing at one student's method, which is indicated by the vertical line band in Figure 1 (proposing phase). Thereafter, the students worked individually on the question and several students presented their ideas. Once again, Mr. Taka focused on their ideas and the discussion continued for approximately 15 to 20 minutes; this is depicted by the dark band (focusing phase). At the end of the lesson, a short summary thereof was presented (formulating phase).

Time (min)	L1	L2
0		
5		Reviewing
10		
15		
20	Working individually	
25		Working individually
30		
35		
40		
45		
50		

Proposing the problem
Eliciting Students' ideas
Focusing on the object of examination
Formulating the result

Figure 1. The four phases in L1 and L2

In L2, the class reviewed the previous lesson and continued their discussion about comparing the three fractions. In L2, they discussed the comparison between 2/3 and 3/4 mainly. To compare these fractions, the eliciting phase and focusing phase were repeated. Interestingly, the formulating phase was observed in the middle of the focusing phase; the students needed to make reference to the previous discussion and accordingly, Mr. Taka summarized what had materialized briefly. Then Mr. Taka proposed a question that could be connected with the previous focused discussion; the students first worked individually, then presented their ideas and finally, engaged in the discussion by focusing on the object of examination.

One notable observation in Figure 1 is the length of time that was devoted to the focusing discussion. In each lesson, approximately 30 minutes of the class hour was spent on the discussion with a focused object of examination. In the next section the content of the students' discussion and Mr. Taka's reflections on the discussion of the focusing phase are discussed.

4.3 *The teacher's role in the focusing discussion in Lesson 1*

Explaining 2/4 < 2/3; the first focusing phase. In this discussion, the students explained why 2/4 < 2/3. They began to talk about unit fractions, which Mr. Taka wanted them to pay attention to, during the focusing discussion. The students used their own vocabulary for "area of one part" and clarified what it means in interactive ways.

The initial interactions between the students and Mr. Taka are thus presented. These interactions reveal how their focus on the area of one part emerged.

> 01 Nishi: The numerators are the same. If we compare two [parts] of what we divided into four and two [parts] or what we divided into three, 2/3 is larger.

02 Mr. Taka: [After writing Nishi's explanation on the blackboard] Is there anyone who can explain this in more detail?

03 Ino: The one that is divided into three [is larger], because the area of one part is larger, so we know 2/3 is larger (In Figure 2, a record of her utterance is depicted by Mr. Taka).

04 Naka: [I can explain it] in detail. Mine is in detail!

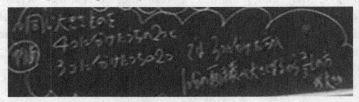

Figure 2. An explanation by Ino, which was recorded on the blackboard

05 Mr. Taka: Alright. So, you can explain what Ino said in more detail, can't you?

06 Naka: Yes, yes. [He walked in front of the blackboard and began to draw cups.] I used a figure. Two of the four equal parts are this part, and ... 2/3L means, well, divide this into three equal parts, and take two of them, they are here. [Mr. Taka added slanted lines.] So it means that 2/3L is [larger]... (see Figure 3).

Figure 3. Two cups by Naka

07 Mr. Taka: Is what you are saying the same as what Ino has just said? I understand your explanation itself very well. Very good explanation.

08 Student: [Naka said] "area." I think that is different [from what Ino said.]

09 Naka: I did not say area.

10 Mr. Taka: What Ino said was, the same amount was divided by four, she divided it into four, and then took two of them. This is what Naka did [Mr. Taka added lines in the left cup as shown in Figure 4]. And the amount was divided by three and took ... [Mr. Taka added lines in the right cup.]

11 Ino: Naka's figure is different [from what I said].

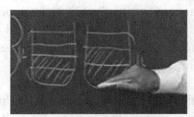

Figure 4. Addition by Mr. Taka

12 Mr. Taka: The figure is different? You [Ino] said, take two of three, so this part is missing. [Mr. Taka underlined Ino's utterance, which was recorded on the blackboard with yellow chalk. This is shown in Figure 5. It reads, "area of one part."]

Figure 5. Underlining Ino's explanation

13 Naka: What? Area? ... [Looking at the underlined part], but this is "Litre."

14 Mr. Taka: It says "area of one part." [To Naka,] What does it say about Litre?

15 Hana: No, it isn't. The area of one part is this part. [She walked in front of the blackboard and highlighted the parts 1/4 and 1/3 in Figure 4.]

The informal language, "area of one part" was first uttered by Ino (Line 03). Mr. Taka recorded her explanation on the blackboard. Then Naka volunteered to develop a figural representation. It was about this figure that Mr. Taka posed the question of whether it really represented what Ino said (07). Mr. Taka checked carefully which part of the figure represented which part of Ino's explanation (10, 12). During this period, Ino mumbled that Naka's figure might not represent what she had said (11). Mr. Taka drew the students' attention to the missing part of Naka's figure, which was "area of one part" (12). Naka appeared to be perplexed and tried to clarify what was meant by "area of one part" (13, 14). Another student, Hana, contributed to the discussion by making the meaning clear by using red chalk (15).

This revealed that the student's explanation had been refined and there was a clearer focus on the concept of unit fractions; this was mediated by their informal language. A driving force of this progression of interaction lay in Mr. Taka's question about a student's explanation; he asked whether someone could add to her explanation. Furthermore, he asked if what Naka drew represented what Ino had said. By directing the students' attention to checking if their discussion was characterized by logical consistency, they paid more explicit attention to the unit fraction.

Explaining why they need 12; the second focusing phase. In this discussion, the students explained why they need 12 to compare 2/4 and 2/3. The least common multiple 12 was first suggested by Miku when she explained the reason for 2/4 < 2/3 in the first focusing phase:

16 Miku: It is easier to know the difference with mine. The least common multiple of 3 and 4 is 12. So, I divided a rectangle into 12. I connected 12 blocks. This is one block [pointing to 1/12 in Figure 6]. [Mr. Taka marked it with red chalk and wrote "a block" in Figure 6.] For 2/4, I divided the blocks into 4 chunks, and 1, 2 [counting the chunks], well, I made a mark here [pointing at the area of 2/4, Figure 6]. [She explained 2/3 in the same way.] Then we know that 2/3 is larger by the difference of the 2 blocks.

When the discussion on the first focusing phase had finished, Mr. Taka went back to Miku's explanation and asked the students:

17 Mr. Taka: I really don't understand where this 12 comes from. [He underlined a part of her recorded explanation with red chalk, which was written as "since 12 is the least common multiple of 3 and 4, I divided a rectangle into 12."]

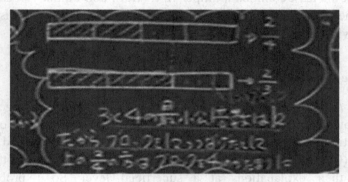

Figure 6. The record of Miku's explanation

18 Students: What? Well... [Some students raise hands.]

19 Mr. Taka: Some of you are raising hands. Okay. So please write the reason you used 12 in your notebook so that everyone can understand easily. I will give you five or six minutes.

After working individually, several students gave their reasons:

20 Sei: We must make the denominators of 2/4 and 2/3 the same number. ... We cannot compare them unless we multiply the same number. ... Well, [if we multiply] 3 and 4, they become the same number. 12 is the number both can be divided into [both 4 and 3] evenly, so we divide 12 by them [meaning 4 and 3].

21 Naka: ... We cannot compare 2/4 and 2/3 at a glance. It is true that we cannot compare them at a glance. But well, if we divide [12] by 4, as she just did, if we divide it by 4 and by 3, we know [the difference is] 2 easily.

Sei repeated the computational procedure (20). Naka insisted that the figure clearly showed the difference (21). Since Mr. Taka further asked the "reason" for the necessity of 12, the discussion continued. Another student, Ida, showed that 2/4, 4/8 and 6/12 represent the same amount by using a tape diagram. The lesson ended. Mr. Taka summarized the discussion by saying, "So you are saying that as a result, you got the least common multiple 12 ... But I don't think you explained how the least common multiple was derived. Not yet." The discussion was reviewed and continued in L2.

4.4 *The teacher's intentions in the interactions in Lesson 1*

Trying to elicit important mathematics in the student's language. In the stimulated-recall interview, Mr. Taka said that the period of these interactions was an important teaching moment. He reflected on his action (02) and said, "I stick to making a common unit... I do not want to finish their explanations at a superficial level. So I probed her explanation by saying 'in more detail'." With reference to Ino's explanation (03), he said, "Ino (03) was on target because she talked about the reason one can compare the fractions of the same numerator in relation to area.... The reason she gave may be linked to the notion that the fractions can be compared by the amount of the unit fraction. So this is very important." With reference to Naka's explanation (06), he added that it was within his expectation and that if no one responded, he was going to ask the students if there was anyone who had developed their own figures. Mr. Taka's comments indicated that he consistently noticed the importance of mathematics in the student's language and tried to elicit it in his interactions with his students.

Making students realize that they need to think about the meaning. Mr. Taka revealed that he was not hesitant to find out what Miku was thinking (16). He said, "Miku is one of the students who is not good at mathematics. When I looked at her raising hand, I asked her. Every student is precious and I do not want to deal only with the opinions that are convenient to me. But when she began to explain 12, I thought, 'Oh, I should not have asked her.' Everyone knew what she would explain." Mr. Taka emphasized that the students know the procedure of finding the least common multiple, but they probably cannot answer the question why they need to find it. Therefore, he decided to probe them for the reason (17). He said that even though he first thought that he should not have asked Miku, it resulted in an important opportunity for the students to think about the reason.

It is notable that Mr. Taka was checking whether the students were able to make sense of the idea of finding common unit fractions when comparing two fractions. He was especially sensitive to whether the students paid attention to meaning and quantity. In the interview, by reflecting on different opinions proposed by the students (20, 21), Mr. Taka said that by repeatedly asking the reason for 12, he wanted them to verbalize that they need to divide the whole into 12 equal parts to "make the areas of one part the same size." He said, "[what they were doing] was dividing a whole into 12 parts so that each piece is a 1/12. The students probably know it implicitly, but they cannot verbalize it in words."

4.5 *The teacher's role in the focusing discussion in Lesson 2*

Explaining whether a method relates to what they had discussed; the first focusing phase. In L2, the class dealt with the comparison between 3/4 and 2/3. In this discussion, Maya explained that 3/4 > 2/3 by attending to the difference between each fraction and 1:

22 Maya: ... For 3/4, if we add 1/4 it becomes one, and for 2/3, if we add 1/3 it becomes one. I compared the remainders, 1/4 and 1/3. Then, as Naka said earlier, the area of 1/3 is larger, 1/3 is larger, well, so, 3/4 is larger because its remainder is smaller.

By employing this non-standard method, Mr. Taka asked a question and the students discussed whether this method was different from their ways of comparing two fractions by making either the numerators or denominators the same, which they had been using. A transcript showing part of their interactions follows.

23 Mr. Taka: So far, the fractions were able to be compared if either numerators or denominators are the same. This time, [neither numerators nor denominators are same,] so I am wondering whether this method of using 1/4 and 1/3 has nothing to do with the one over there. [He pointed at a recorded summary of the ways they had discussed. It says 'two fractions can be compared if either numerators or denominators are same.']

24 Miya: I think this method has nothing to do with the previous ones. The other method is an easy one because either the numerators or denominators are same, but this one is about the difference.

25 Mr. Taka: Yes, but after that, this method compares 1/4 and 1/3, doesn't it? [It compares] which result is larger. So it compares the remaining amount, 1/4 and 1/3.

26 Miya: I think it probably has something to do with... Making the denominator or numerator the same and add to make 1 and... well...

27 Mr. Taka: Sorry, my question is whether comparing these, namely, 1/4 and 1/3 has a relationship with this. [He pointed at the recorded summary again.] Yes, Sei.

28 Sei: 1/4 and 1/3 have the same numerator. So it has a relationship with the fact that we can compare fractions if either the numerator or the denominator is common.

The interaction demonstrates that at first the student did not notice the relationship between the method and their summary of comparing fractions (24). However, after Mr. Taka's prompt (25), several students recognized the similarity between the two ways. Sei gave a clearer reason (28).

Explaining a method to make numerators the same; the second focusing phase. After discussing Maya's method, Mr. Taka asked the students to produce other methods to compare 2/3 and 3/4. However, on this occasion, he specified the ways of finding a common numerator or denominator. After working individually for about six minutes, two students presented their solutions. One student converted 2/3 and 3/4 into decimal numbers. Mr. Taka asked the students whether someone wanted to add to this method. Since no one raised a hand, he went to the next proposal. The next student, Ino, compared 6/9 and 6/8; this method of finding a common numerator was examined in the second focusing phase.

29 Ino: I made the numerators same. I made the numerators 6 because I thought that the smallest number which can be divided by both the numerator 3 in 3/4 and the numerator 2 in 2/3 is a good one. To change 3/4 to 6/□, 3×2 is 6, and since they are proportional, I did 4×2.

30 Mr. Taka: Wait. Did you think they are proportional?

31 Ino: I thought the numerator and denominator are proportional. Well, I multiplied the numerator by 2, so I thought I needed to multiply the denominator by 2. Then I got 6/8. I did the same thing for 2/3, and got 6/9. Well, well so 6/9..., is 6/8 larger?

32 Mr. Taka: [Is 6/8] larger?

33 Ino: I think so.

With reference to Ino's method, Mr. Taka posed a question about proportional:

34 Mr. Taka: You thought since you multiplied the numerator 3 by 2, you must multiply the denominator by 2. Why? You said "proportional." Can someone explain this so that everyone can understand? ... It may be better if you draw a figure.

In response to Mr. Taka's question, Kari explained his method by drawing figures. He explained the proportional relationship between the numerator and denominator by employing the reasoning of what would happen if he did not multiply the numerator and denominator with the same number. Some of their interactions follow.

35 Kari: [Mr. Taka drew three circles and arrows by copying from Kari's notebook.] I used the method of making denominators the same. This figure shows 3/4 (Figure 7a). First, I only multiplied the denominator by 2, and drew a figure (see Figure 7b). I divided the circle into 8 parts and [shaded] 3 of them. These are the original lines [he colored the two crossed lines with red], and they show that at first the amount is this [pointing at 3/4], but when I multiply the denominator by 2, the amount is less and this [pointing at 3/8]. So we know that the amount has changed.

36 Kari: So, I multiplied both the numerator and denominator by 2, so I got 6/8, and now I got the same size as the original one (Figure 7c). So we know that they are proportional.

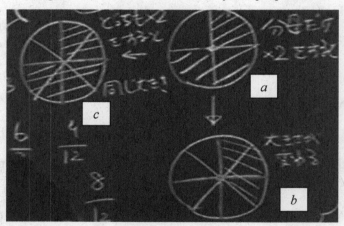

Figure 7. Circles that show an argument by Kari

37 Mr. Taka: So are you saying that you must make them proportional?

38 Kari: Yes, we must make them proportional.

4.6 *The teacher's intentions in the interactions in Lesson 2*

Calling a student intentionally. In the interview, Mr. Taka commented on the first focusing phase as an important teaching moment in two ways. Firstly, it was important because he dealt with Maya's reasoning. In L1, Maya expressed her opinion that 3/4 is larger than 2/3 when Mr. Taka first posed an open-ended question. However, Mr. Taka postponed Maya's reasoning by stressing "at a glance" in his question and said, "I will come back to you later." In the interview, Mr. Taka said, "I remembered that I postponed Maya's thinking in L1, so I called her here... I don't break my promises. It probably is about trust between the students and me. It is about my message to the students." He added, "[Dealing with Maya's thinking is] important in the sense that lessons are connected unavoidably."

Secondly, Mr. Taka expressed his discouragement when he knew that the students did not easily understand the similarities between Maya's method and the summarized ways of comparing fractions. Furthermore, he said, "But later, they found that they were the same. This part was a little bit interesting because there was a conflict of opinions... My way of asking questions might not be good..."

As these comments suggest, Mr. Taka called a student intentionally for pedagogical reasons. It is also noteworthy that he valued the connections between the lessons. Consequently, the class naturally adopted a non-standard method. As a result of Mr. Taka's questions on the connections of the unique method with the previously shared method, the students paid attention to the ideas they were learning.

Trying to build on the students' thinking toward the lesson goal. In L2, the class discussed a non-standard method again, which was developed by a student. The interview with Mr. Taka revealed that he called Ino intentionally. He said, "When I was walking around the desks [during the individual work], I found that Ino was making the numerators the same. I thought I would call her if she raised her hand." He spoke about the reason: "I wanted to first listen to the opinion of making the numerators same. I did not want to suggest the method of making denominators the same first because it would direct the students' attention to calculation." During the time the students worked individually, Mr. Taka also noticed that several students were thinking about the reasons for working out the least common multiple. He said, "I thought I would engage these students later in the discussion."

With reference to Ino's presentation, Mr. Taka probed her by saying, "proportional" (30). In the interview, he commented that this instruction was important because "it will clarify the meaning of finding the least common multiple." Mr. Taka was dissatisfied with the students' view of proportional relationships as though such relationships had existed previously. He was making sure with Kari that multiplying both the numerator and denominator by 2 or 3 *resulted in* equivalent fractions and therefore, that *it was up to us* to make them proportional (37 and 38). Mr. Taka was not satisfied with Kari's explanation either. He commented on Figure 7b, "It shows 3/8 by multiplying only the denominator by 2. But multiplying by 2 for denominator really means making its scale smaller. So 3/8 never happens. Since we reduced the scale, one measure became two measures or [4] into 8, it automatically results in 6. This is what is really happening. But the students do not think so. They pay attention on to how to calculate the least common multiple."

The interview indicated that Mr. Taka was trying to elicit the students' reasons for the important points he wanted to teach and to build on the students' thinking so as to achieve the goal of the lesson.

5 Conclusion

In this chapter, the roles of goal of the lesson that a teacher has in mind so as to control classroom interactions were explored. By examining two consecutive lessons conducted by an experienced teacher, the nature of the goal of a lesson goal that the teacher set as well as several functions of the goal to construct classroom interactions and to manage different reactions from the students were revealed.

One notable observation is that the teacher possessed a clear goal of the lesson and moreover, a network of goals that encompassed the whole unit of study. The teacher had a firm understanding of how important it was for the students to learn units and unit fractions in L1 and L2 so as to understand both the meaning and procedure of the addition and subtraction of fractions with unlike denominators.

It is clear that the teacher's goal of the lesson had a profound influence on his construction of interactions and management of a variety of students' responses. All four focusing phases described in the previous section were concerned with unit fractions; these were linked to the goals the teacher had in mind. He was sensitive to the students' language, written representations and the order of writing symbols and accordingly, he elicited their understanding of unit fractions. He often highlighted the gaps between their understanding and what he intended and asked for clarification when he was unsure of the language they had used as well as for justifying a non-standard method. It was also clearly shown that the teacher resisted the students' inclination toward symbolic manipulation without thinking about the meaning thereof. The value he placed on the concept of unit fractions provided a foundation for his teaching methods.

Many of the teaching actions and interactions described in this study are consistent with those identified in the previous studies. In particular, this study found a sub-system of interaction in which the students interpret and clarify simple language or solutions by their peers so that everyone can understand. In L1, they interpreted and clarified the meaning of the area of one part. In L2, they interpreted non-standard methods by their peers and clarified its connection with their shared method. The driving force of this interaction is trying to understand what the peers are thinking and proposing. Therefore, dominant moves of interaction between the students add to the previous explanation, modify it or give it a new interpretation. The teacher led this interaction by asking questions and furthermore, by triggering conflict and perplexity. In the interview, Mr. Taka said, "When I found that the student's response was incomplete, I thought, 'Oh, I am lucky,' because by elaborating on it, they can focus on the unit fraction." This comment shows that the teacher valued such interaction in developing the students' understanding.

It is also notable that the pedagogical value of interpersonal relations ensured a solid foundation for Mr. Taka's teaching actions and interactional moves. For him, establishing the relational basics with the students should be prioritized before anything else. He made decisions on who he should ask to do a presentation on the basis of mutuality of trust. Because of this first priority, he experienced a difficulty from the perspective of his lesson goal. Nevertheless, he was responsive to the students' opinions and made flexible on-the-spot decisions. The *relational framework* postulated by Sato (1996), in which connectedness is the central idea, captured a feature of the teacher's classroom practice.

In this study, the case of an experienced teacher in Japan was examined. Although the results obtained by the analysis are neither typical nor generalizable, they offer information on the role of the lesson goal the teacher had in mind to construct and manage classroom interactions. Hill, Rowan, and Ball (2005) claim that teachers who are able to develop students' mathematical understanding are those who have a sound foundation of subject knowledge. This study adds to their claim by revealing that teachers also have a clear and rich goal-related subject knowledge and solid educational vision for teaching the subject.

Acknowledgment

The author thanks Yuka Funahashi for her helpful comments in an earlier version of this chapter.

References

Chazan, D., & Ball, D. (1999). Beyond being told not to tell. *For the Learning of Mathematics, 19*(2), 2-10.

Clarke, D. (2006). The LPS research design. In D. Clarke, C. Keitel, & Y. Shimizu (Eds.), *Mathematics classrooms in twelve countries: The insider's perspective* (pp. 15-36). Rotterdam, The Netherlands: Sense Publishers.

Fujii, T. (2013). *Cross-cultural studies on "collective thinking" in mathematics "lesson study" among U.S., Australia and Japan.* Research Report of Grants-in-Aid for Scientific Research by Japan Society for the Promotion of Science (No. 20243039), Tokyo Gakugei University (in Japanese).

Fujii, T. (2015). The critical role of task design in lesson study. In A. Watson & M. Ohtani (Eds.), *Task design in mathematics education: An ICMI Study 22* (pp. 273-286). New York: Springer.

Funahashi, Y., & Hino, K. (2014). The teacher's role in guiding children's mathematical ideas toward meeting lesson objectives. *ZDM Mathematics Education, 46*, 423-436.

Hill, H., Rowan, B., & Ball, D. (2005). Effects of teachers' mathematical knowledge for teaching on student achievement. *American Educational Research Journal, 42*, 371-406.

Hino, K., & Koizumi, Y. (2014). Interactive construction of attention paths toward new mathematical content: Analysis of a primary mathematics lesson. In S. Oesterle, P. Liljedahl, C. Nicol, & D. Allan (Eds.), *The 38th Conference of the International Group for the Psychology of Mathematics Education and the 36th Conference of the North American Chapter of the Psychology of Mathematics Education* (Vol. 3, pp. 305-312), Vancouver, Canada: PME.

Lewis, C., & Hurd, J. (2011). *Lesson study step by step: How teacher learning communities improve instruction.* Portsmouth, NH: Heinemann.

Lobato, J., Clarke, D., & Ellis, A. B. (2005). Initiating and eliciting in teaching: A reformulation of telling. *Journal for Research in Mathematics Education, 36*(2), 101-136.

Lobato, J., Hohensee, C., & Rhodehamel, B. (2013). Students' mathematical noticing. *Journal for Research in Mathematics Education, 44*(5), 809-850.

Sato, N. (1996). Honoring the individual. In T. P. Rohlen, & G. K. LeTendre (Eds.). *Teaching and learning in Japan.* Cambridge (pp. 119-153). UK: Cambridge University Press.

Selling, S. K. (2016). Making mathematical practices explicit in urban middle and high school mathematics classrooms. *Journal for Research in Mathematics Education, 47*(5), 505-551.

Shimizu, Y. (Ed.). (2011). *Cross-cultural studies of mathematics classrooms from the learners' perspective.* Research Report of Grants-in-Aid for Scientific Research by Japan Society for the Promotion of Science (No. 19330196), University of Tsukuba (in Japanese).

Walshaw, M., & Anthony, G. (2008). The teacher's role in classroom discourse: A review of recent research into mathematics classrooms. *Review of Educational Research, 78*(3), 516-551.

Wood, T., Williams, G., & McNeal, B. (2006). Children's mathematical thinking in different classroom cultures. *Journal for Research in Mathematics Education, 37*(3), 222-255.

Chapter 9

Designing and Implementing Activities in the Flipped Classroom in the Singapore Primary Mathematics Classroom

CHENG Lu Pien   NG Swee Fong
TAN Bee Kian Jasmine Susie   NG Ee Noch

The flipped classroom approach has been extensively studied in the context of higher education as compared to the elementary school. In this chapter, we describe a model of flipped classroom approach adopted by some primary mathematics teachers in Singapore. We discuss the benefits as well as some of the challenges the teachers faced in their flipped classrooms. The teachers' lived experiences in the implementation of the flipped classroom approach provides possible guidelines on the design of the flipped classroom in primary mathematics classrooms.

1 Introduction

Problem solving is central to the Singapore mathematics curriculum. However, some learners face problems with retention and retrieval of mathematical skills, procedures and concepts. If the bulk of classroom instruction time is spent mainly on helping students acquire and recall basic mathematical facts, skills and procedures, then the opportunities to develop mathematical reasoning and thinking skills during problem solving may be limited. One typical primary school in Singapore adopted the flipped classroom with the intention of freeing more classroom time for mathematical problem solving and at the same time facilitate students' learning and recollection of the basic mathematical facts, skills and procedures. The first author met one of the teachers, Teacher Linda, from the school during a professional sharing session and both found common interest in examining how the flipped classroom can enhance teaching and learning mathematics in the primary mathematics classrooms. The first author then invited Teacher Linda and her school to participate in research on flipped classrooms. Two teachers played a key role in this study, Linda and James. Henceforth they are identified as Teacher-L and Teacher-J respectively.

This chapter reports how the flipped classroom can look like in primary mathematics classrooms in Singapore using phenomenological approach, relying on teachers' accounts of their experiences in the flipped classrooms. Specifically (i) What do mathematics teachers actually do in a flipped classroom? (ii) What are the experiences of teaching and learning in a flipped classroom?

2 The Model of Flipped Classroom Approach in this Study

According to Bergmann and Sams (as cited in Lai & Hwang, 2016) the term *flipped classroom* represents a learning approach that "exchanges the time used to deliver basic knowledge in class and the out-of-class time for applying the knowledge or doing homework" (p. 127). This means that during the mathematics lessons, teachers would be able to engage students more in activities that require them to apply the knowledge they learned. Learning materials for out-of-class learning could be related to knowledge levels of remembering and understanding e.g. instructional videos, web-based tutorials (Lai & Hwang, 2016).

One theoretical argument in the use of the flipped classroom is that it has the potential to help manage students' cognitive load and improve students' motivation through two pedagogical theories: self-determination theory and cognitive load theory (Abeysekera & Dawson, 2015). In Bishop and Verleger's (2013) discussion on the theoretical frameworks used to guide the design of in-class activities, they pointed out that the theoretical foundations used to justify the flipped classroom usually focus on "reasons for not using classroom time to deliver lectures" (p. 5). Such reasons stem from the large corpus of literature on student-centered learning supported by student-centered learning theories. Another argument for the flipped classroom approach is to develop students to be self-directed learners. For example, a self-regulated learning system was developed in the flipped classroom approach for fourth grade students at an elementary school for their out-of-class learning (Lai & Hwang, 2016). The system allowed students to take quizzes before in-class activities, set their learning goals and track their learning performance. Their teachers could also provide feedback on students' learning processes through the system. Self-regulation is important especially for out-of-class learning in the flipped classroom approach as the activities in the in-class learning draw upon the learning acquired in the out-of-class learning.

The flipped classroom approach used in this study adopted the definition by Bishop and Verleger (2013). They defined the flipped classroom as "an educational technique that consists of two parts: interactive group learning activities inside the classroom, and direct computer-based individual instruction [include only designs that employ videos] outside the classroom" (p. 5). Using Bishop and Verleger's definition of the flipped classroom approach, a systematic review was conducted by Lo and Hew (2017) to investigate the implementation of K-12 flipped classroom. Their review provided very useful information on the practice of flipped classroom in K-12 education, such as, the effects of flipped classroom approach on K-12 student achievement, K-12 students' attitude towards the approach and challenges of using flipped classroom in K-12 education. We used the literature on flipped classroom to make explicit the model of flipped classroom approach used in this study to address the problem of retention and to create more opportunities to develop mathematical reasoning and thinking through mathematical problem solving. The model used in this study addresses three considerations; (i) *meaningful mathematical discovery or exploration,* (ii) *use of technology* and (iii) *apply concepts and skills to problem solving.*

Meaningful mathematical discovery or exploration. Indeed, the problem of retaining once-learned material can be a serious problem faced by many students (Bahrick & Hall, 1991). Students either forget quickly a large portion of knowledge they have learnt in school or they cannot retrieve the knowledge later on, that is, students have difficulty retaining the knowledge. Memory plays a significant role in knowledge retention. When students participate in the classroom experiences, internalization of the classroom activities will help students acquire memory systems e.g. episodic memory system (Nuthall, 2000). Thus the activities in classrooms are vital because they can be an aid to help students remember what was taught (Engelbrecht, Harding, & Preez, 2007). The literature suggested that one of the ways to improve retention is through different instructional approaches, for example, active learning approaches, students' involvement in inquiry and discovery and scientific processes improves knowledge retention (Handelsman et al., 2004). Also, students were reported to exhibit higher retention of knowledge via experiential learning as compared to students engaged in a traditional lecture-style manner (Van Eynde & Spencer, 1988). Meaningful mathematical discovery or exploration if coupled with concrete materials may further aid students' learning. Indeed, "the appreciation of mathematical ideas through concrete constructions using materials of various kinds for these constructions" (Bruner & Kenney, 1965, p. 51) is one of the steps towards mathematical learning. Furthermore, "A common thread among the three theorists [Bruner, Piaget, Dienes] is an existence of parallel stages, analogously sequenced from concrete to abstract ..." (Chang, Lee, & Koay, 2017, p. 4). According to Bruner & Kenney (as cited in Chang, Lee & Koay, 2017), from the constructions, students can then form "perceptual images of the mathematical idea" (p. 5) and develop a symbol to describe the constructions. This approach aligns with the Concrete-Pictorial-Abstract instructional sequence (with its roots from Bruner (1964) for conceptual learning) for the development of primary mathematics concepts in Singapore. Teaching for conceptual understanding can lead to longer retention of mathematical knowledge (Kwon, Allen, & Rasmussen, 2005).

Use of technology. "Research suggests that enabling technologies have the potential to increase engagement and retention if implemented in the correct manner" (Wankel & Blessinger, 2013, p.15). Different types of enabling technologies are used in education e.g. student created videos

(Engin, 2014), mobile-assisted learning system (Wang, 2016), YouTube and Khan Academy Podcast. "The internet and advances in streaming video have greatly improved students' access to information" (Davies, Dean, & Ball, 2013, p. 566). The use of videos in teaching is not new but videos can have such a powerful effect on one's mind and sense that "you may download it off the internet or order the DVD from Amazon along with the CD soundtracks so you can relive the entire experience over and over again" (Berk, 2009, p. 2). Drawing on multimedia and cognitive learning theory, Berk (2009) summarised pertinent suggestions on the use of videos for students to experience the "powerful cognitive and emotional impact" (p. 2) provided by videos. Some of the suggestions include techniques for using video clips in teaching and the selection of appropriate videos for classroom use. However, it is "not an easy task to find videos that perfectly match what a teacher wants his or her students to learn" (Chen, 2016, p. 418). As a result, teachers may also create their own videos, also known as crafted video lessons. Care and expertise in crafting video lessons is necessary and Huang (2016) developed a framework using Gagné and Brigg's (1979) nine events of instructions to guide the crafting of video lessons incorporating multi-modal representations to enhance relational understanding of the factorization concept. Content notes can be given to students to guide their note taking when viewing the videos (DeSantis, Van Curen, Putsch, & Metzger, 2015). They, however, cautioned against using content videos as a replacement instead of enhancement of classroom instruction and further added that "this misuse of flipped lesson planning can negatively affect the quality of instruction delivered by teachers who adopt the model" (DeSantis et al., 2015, p. 44). Furthermore, the power of the flipped classroom does not lie in the videos, rather, it lies in how teachers spend the instructional time with their students (Bergmann & Sams, 2012).

Apply mathematics concepts and skills to problem solving. The flipped classroom approach has the potential to engage students in higher-order thinking rather than receiving direct teaching instruction (Davies, Dean, & Ball, 2013). Higher-order thinking includes different types of thinking. According to King, Goodson, & Rohani (as cited in Daher, Tabaja-Kidan, & Gierdien, 2017) "higher-order thinking includes critical, logical, reflective, metacognitive and creative thinking" (p. 1). It is necessary to cultivate higher-order thinking in our students as such thinking empower them to be more critical thinkers thus supporting them in problem solving in mathematics.

In this study, the three considerations above guide the design of the flipped classroom model in the teaching of one chapter from the textbook. Some of the activities in the flipped classroom in this study are similar to Lo and Hew's (2017) model of flipped classroom approach. Figure 1 shows the flipped classroom model used by the teachers in this study.

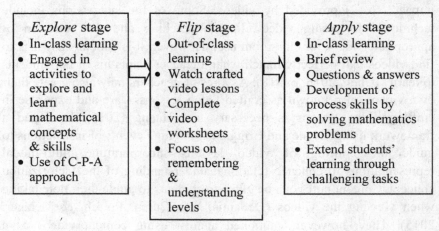

Figure 1. Model of flipped classroom approach used by the teachers

Explore stage. In-class activities and concrete materials are used usually at this stage, usually at the beginning of a mathematics topic to engage students in mathematical inquiry in order to develop students' relational understanding "knowing what to do and how" (Skemp, 2006, p. 89) of mathematical concepts.

Flip stage. The second stage involves out-of-class activity whereby students view assigned crafted video lessons (on-line). Similar to Lo and Hew (2017), the central teaching strategy in the videos is direct instruction focusing on knowledge levels of remembering and understanding. For example, the videos can be explanation of a mathematical skill, straightforward methods and computations, examples to elaborate concepts (introduced in *Explore* stage). The crafted video lessons can also be used as a consolidation tool and/or in preparation for the next mathematics lesson. This taps on the affordances of crafted video lessons in providing the additional assistance to aid students' remembering of basic skills and procedures. Accompanied worksheets and reading materials were designed to assist students' learning when they view the crafted video lessons. The worksheets contain questions for students to write down their responses.

Apply stage. In-class activities in this stage are similar to Lo and Hew's (2017) model. A brief review on the crafted video lessons to help students recall materials from the video and to clarify misconceptions. Question-and-answer sessions may also be conducted to clarify any doubts and misconceptions in the crafted video lessons and video worksheets. Students then engage in learning activities to apply knowledge learned from the crafted video lessons and to solve advanced problems. Similar to Lo and Hew, the teachers can also extend students' knowledge or explain difficult or complicated concepts at this stage. According to Carroll (as cited in Eustace, 1969) complex concepts are characterised by "their hierarchical nature; that is, their dependency on a network of related or prerequisite concepts" (p. 449).

2.1 *An example of the activities in the flipped classroom approach*

One of the topics taught by the teachers using the flipped classroom approach was triangles at Primary 5. There are altogether four lessons using the flipped classroom approach for the topic on triangles. In the *Explore* stage (Lesson 1), the teachers engaged the students in proving angle properties such as angles on a straight line add up to 180°, angle sum of triangle is 180° and angles at a point is 360°. Teacher-J said that:

> For this particular topic, our main aim was to get the kids to prove because there are a few … like angles on a line and at a point, these

are a few things that they have to remember. So instead of us just telling them that this is angle on a line, remember it. So what we did was we got them to explore to do hands on, to prove for themselves that angles on a line is 180°.

For example, to prove that sum of angles in a triangle is 180°, students can be given different types of triangles (isosceles, equilateral, right angled, obtuse-angled, acute-angled etc.) as shown in Figure 2 to cut out. They can tear the three angles in each of the triangles and arrange them on a ruler to arrive at the sum of the angles of a triangle is 180°. This activity was adapted from the textbook. Teacher-L added that the *Explore* stage is particularly useful when a new concept is introduced "if it is an introduction to a new concept, a new topic, so we would do an explore part —where they will explore first."

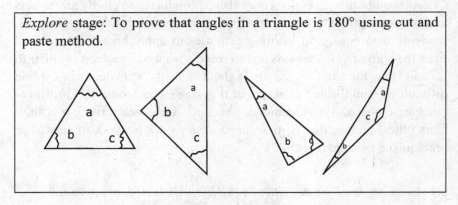

Explore stage: To prove that angles in a triangle is 180° using cut and paste method.

Figure 2. An example of an activity in *Explore* stage

Following the *Explore* stage, is the *Flip* stage. As all the crafted videos for all the Primary 5 mathematics topics were housed under one website, the mathematics teacher will need to inform the students the specific crafted video (related to the hands-on stage for students) to view after school. According to Teacher-L, "So the video, usually our videos is telling them [the] procedure how do you solve angles in a triangle? But they would already have that explore part where they say that angles on a triangle is 180°, angles on a straight line is 180°". Teacher-J added that the crafted videos included various examples and methods for the students to find an unknown angle in a triangle given the other two angles in the triangle. "Similar question, after they watch if they understood, they would solve on their own." (Teacher-J). Figure 3 illustrates a screenshot of the crafted video.

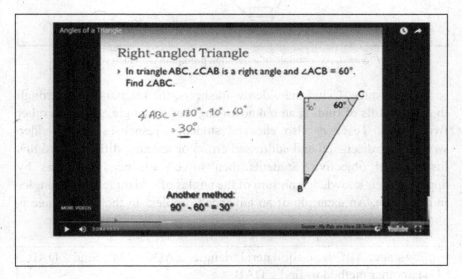

Figure 3. A screenshot from the crafted video in *Flip* stage

Some of the crafted videos include question prompts and the students are required to provide a response for those question prompts in the video worksheets as shown in Figure 4. The video worksheets were necessary as they helped students to focus and take note of key learning points in the crafted videos and identify any learning difficulties they had after viewing the crafted videos.

Some of the video worksheets contained reflection prompts such as "Reflect and write your questions below" to engage students in self-reflection of their learning acquired from viewing the crafted video lessons. Such prompts also allow students to communicate to their mathematics teachers their thoughts and feelings about the crafted video lessons and the mathematics presented in the videos.

Flip stage: Watch the video to fill in the missing angles. Find the unknown angle ∠MLN.

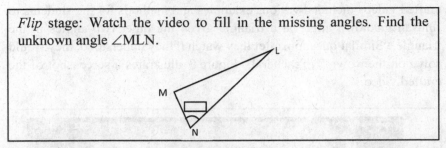

Figure 4. An example of a question prompt in the video worksheet

Apply stage. Using the video worksheets, the teachers went through the basic skills of finding an unknown angle of a triangle, given the other two angles. Teachers also checked students' responses to the video worksheets, discussed and addressed errors or learning difficulties to this instructional objective. Students then solved advanced problems by applying their knowledge on sum of the angles of a triangle and the angles in a triangle. An example of an advanced problem in the *Apply* stage is shown in Figure 5.

Apply stage: ABC is an equilateral triangle. ∠ADC is 98°. Find ∠DAB. Use another method to find ∠DAB.

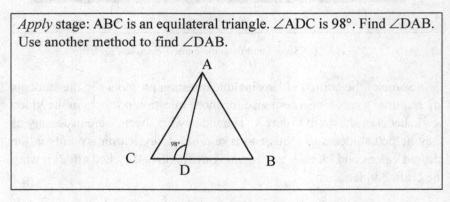

Figure 5. An example of a problem in *Apply* stage

Teacher-L and Teacher-J implemented the A*pply* stage very differently (Lesson 3 to 4). In Teacher-L's mathematics classroom, the students mostly worked individually when solving mathematics problems with Teacher-L attending to students that encountered obstacles at different stage of the problem solving process. Group work was used only occasionally. In Teacher-J's mathematics classrooms, students solved mathematics problems mostly in groups of four. The mixed-ability grouping provided more opportunities for students to engage in collaborative learning as well as self-directed learning. Students were encouraged to check with their team or seek help from their team members when they faced any obstacles in the problem solving process before seeking the teachers' help.

According to Teacher-J, this approach not only encouraged collaborative learning, it also encouraged those who were able to solve the problem to explain and communicate mathematical ideas, strategies, and procedures to students who struggled with the mathematical problem solving process. In addition, more time to solve mathematics problems in class enabled Teacher-J to address better the learning needs of the low-attainers in his mathematics class.

3 Benefits for Teachers

3.1 *Professional development*

The flipped classroom approach was first adopted by some Primary 5 teachers of the school. Eventually, more teachers in the school showed interest in the flipped classroom and volunteered to be part of the flipped classroom project. With more people in the team, and with the support of the school management, the team produced the entire teaching and learning package for the flipped classroom for the Singapore Primary 5 mathematics curriculum. The package consisted of lesson ideas, crafted video lessons, video worksheets and problem solving worksheets. The team was in the process of fine-tuning the package (e.g. worksheets and crafted videos) for the flipped classrooms when the researchers exited the research field.

The development of the package afforded the team opportunities to re-examine and deepen their understanding of the Singapore primary mathematics curriculum as the flip classroom approach required the team to make adjustments to the mathematics scheme of work provided by commercial instructional materials. Although the conceptualization and design of the package required a tremendous amount of effort by the team, Teacher-L supported the project. She explained that "I find it rewarding in the sense that there is a rich discussion in terms of teaching, professional development. And also, I find that there are a lot of interactions among primary 5 teachers". The rich discussion included areas such as teaching strategies to teach a specific mathematics topic and variation in examples. Teacher-L recalled her engagement in critical reflection in one of the discussions on the choice of triangles: "Why must we stick to isosceles and equilateral …? Because we are learning triangles. The base and height. So we can actually introduce all these [triangles in addition to isosceles and equilateral triangles] and link them together". Teacher-L felt that it was the rich discussion that attracted more teachers to the group. She opined that "So, we had this big discussion, the different types of triangles. So with that we get everybody excited".

For Teacher-L, the discussion and design of the package was important. She said "the pedagogies, the content, the pre-requisites on how we go about teaching a topic is discussed. Everyone in the team would have a standardized format to teach the content of the topic". The discussion was a platform for her and her team to understand explicitly the content that needed to be taught for each topic, establish common goals in the teaching and learning of mathematics and to be very clear how the teaching strategies may look like to achieve the group's common goals. Teacher-L explained:

> To me I think conceptual understanding of a particular topic of the content is important. Using this platform we discussed the content in the crafted video and how we want to approach teaching the concept. So it is conceptual understanding that we are looking at. We also raised common misconceptions and how we can address these either in the video or in the activities.

3.2 *Learning about crafted videos*

The flipped classroom in this study used crafted videos in the *Flip* stage. PowerPoint screencast was used for the videos to provide step-by-step instructions to guide students' viewing of the videos. The crafted videos facilitated learning because the crafted videos were tailor-made with their school students' profile, learning needs and learning styles of the students in their schools in mind. The crafted videos were produced by the teachers themselves. The teachers' reflection on their crafted videos and feedback from their students, was an on-going process as the teachers produced and refined the suite of crafted videos for the flipped classroom. For example, in this study the teachers reflected that each crafted video should adhere to the 10-minute rule as this was the duration that can hold students' motivation in learning before they got bored. Students get disengaged when the instructional videos were long (Kettle, 2013; Medina, 2014; Schultz, Duffield, Rasmussen, & Wageman, 2014). Guo, Kim, and Rubin (2014) suggested that each video should be about 6 minutes. Besides using their voices, some of the crafted videos had the teachers' faces (some teachers were willing to have their faces captured in the crafted videos upon the request of their students) in them and this feature was important to help some of the students connect with the mathematics in the crafted videos. DeSantis et al. (2015) reported that student's satisfaction in their flipped classroom was significantly lower than the "traditional" counterpart when the instructional videos produced by their team members "did not feature the host teacher" (p. 51).

Other advantages of using the crafted videos were students' ability to control and regulate their learning as they were able to rewind, pause and play the crafted video as many times they like. This aligns with what Schultz et al. (2014) reported "most students had a favorable perception about the flipped classroom noting the ability to pause, rewind, and review lecture" (p. 1334). This function helps the students to manage their cognitive load as they were able to learn at their own pace. Students could also make their own notes to help them make sense of the key mathematical ideas as they watched the crafted videos. The provision of such crafted video lessons meant that these students could have access to such videos and watch them on demand anytime, anywhere.

3.3 *Facilitating teaching*

Though re-teaching of some concepts or skills is still necessary, the teachers were also able to make reference to crafted videos - grounded images (Ng, 2015), that students have watched to build upon a lesson or trigger students' recollection of a previously taught concept or skill or to link mathematical ideas or concepts. Students can also recapitulate some of those concepts and skills on their own through the crafted video lessons. Hence, there was less re-teaching and more time for problem solving. Thus, placing the teaching of mathematical skills and procedures onto the crafted videos frees up more class time for discussion on problem solving of mathematics in class. That is, the flipped classroom afforded more opportunities for students to experience high-order thinking processes and more opportunities to develop mathematical reasoning and thinking in the mathematics classrooms. As such, the teachers found themselves co-teaching as the crafted videos complemented their teaching in the actual mathematics classrooms.

4 Benefits of the Flipped Classroom for Students

More interactions with the students during problem solving sessions gave the teachers more opportunities and time to understand their students' needs, assess their students' learning and problem solving abilities, diagnose and remediate students' learning difficulties immediately. Teacher-L explained:

> And we gave immediate feedback – why is it wrong, why it is correct. … Immediately we, I can explain to them. And we do a few sums, quite a fair bit, and their homework is like maybe 3 or 4 sums and those few that are unable to complete. So with that I think when they go back and do … they are able to… and they find achievement in a sense.

Immediate feedback given to students' performance in class thus allowed students time to re-work and fine-tune the reasoning required to solve the mathematics problems. The immediate feedback allowed the students' to experience small bite-sized success and achievement in problem solving, hence, helped build students' confidence to solve mathematics problems. The immediate feedback also allowed students to complete "similar" mathematics problems at home, thus decreasing the chances of non-submission of homework. Opportunities to provide timely assistance in flipped classrooms was also reported in Tsai, Shen, and Lu (2015).

More class time was therefore used to apply the new knowledge to solving problems. This was also reported by Chao, Chen, and Chuang (2015), Mazur, Brown, and Jacobsen (2015) and Schultz at al. (2014). In that sense, Teacher-J explained that "It [viewing the videos out-of-class] is something easy for them to do at home rather than struggling with the work [applying new knowledge in solving problems]". Teacher-J encouraged his high-progress learners to bring their own challenging mathematics problems to class. Once they had completed solving the core set of problems to be solved by the class, these high-progress learners could then proceed to work on the challenging mathematics problems which they themselves were interested to solve. Giving students choice meant that Teacher-J could differentiate the pace of the students in his class.

5 Challenges of the Flipped Classroom

Challenges were also encountered in this study from the teachers' perspectives. For example, some students did not have internet access or computers at home—an operational challenge (Lo and Hew, 2017). When students did not have access to the internet or computers the teachers would arrange for the school's computer labs to be available for students to view the crafted video lessons after school hours. Alternatively, arrangements would be made for students to view the videos before class began the following day. For some of the high-progress students, the flipped classroom required them to work harder because they had to solve more problems in class rather than having more time to "relax and listen" to the teachers' demonstration and explanation during class time.

Some students were not interested in solving mathematics problems and the flip classroom did not appeal to them. Teacher-J revised his flip classroom in response to such challenges. He incorporated 'flipped mastery' into his flipped classroom by designing the problem solving tasks according to the level difficulty. In his *Apply* stage, the students progress from level 1 problems to level 2 problems before proceeding to the most challenging problems, level 3, if they solve the problems successfully at each level. Teacher-J found his students more motivated to solve mathematics problems through the flipped mastery approach.

6 Conclusion

In this study, the two teachers found themselves more efficient in their use of mathematics classroom time because the *Flip* stage reduced re-teaching during classroom time, thus enabling their students to progress to problem solving. The teachers adopted the flip classroom approach because they believed, appreciated, understood and found value in the approach. The teachers understood that each of the three stages in the flip classroom had a unique function. They shared common goals through the use of the flip classroom approach and they were very clear of their responsibilities. As such, they were able to expand the flip classroom to another (the Primary 4) grade level as more teachers became interested in the flip classroom. In addition, the team of teachers that came together were able to identify, harness the expertise of all its members and work in tandem to develop and implement the flip classroom. Such synergies are necessary for innovative pedagogies.

At the time of writing this chapter, the team had completed the second phase of the research on the flip classroom. In the second phase, the team invested more time in refining their package, and re-examined the crafted video lessons using mediation strategies. Moving forward, the team can

(i) embed a tracker in the crafted video lessons to track which specific instances of the crafted videos were 'most watched'. Such a device meant that it was possible to examine what concepts students found challenging. This meant that teachers could review such concepts with the class to ensure that students have better grasp of this particular concept, and (ii) provide stronger theory-practice links in the flipped approach in the primary schools.

Acknowledgement

The authors wish to thank the teachers and the school who participated in the flipped classroom approach whose work formed the crux of this chapter. The findings of this study were reported at the 13th International Congress on Mathematics Education at Hamburg, Germany.

References

Abeysekera, L., & Dawson, P. (2015). Motivation and cognitive load in the flipped classroom: definition, rationale and a call for research, *Higher Education Research & Development, 34*(1), 1-14.

Bahrick, H. P., & Hall, L. K. (1991). Lifetime maintenance of high school mathematics content. *Journal of Experimental Psychology: General, 120*(1), 20-33.

Bergmann, J., & Sams, A. (2012). *Flip your classroom: Reach every student, in every class, every time.* Washington, DC: International Society for Technology in Education.

Berk, R. A. (2009). Multimedia teaching with video clips: TV, movies, YouTube, and mtvU in the college classroom. *International Journal of Technology in Teaching and Learning, 5*(1), 1-21.

Bishop, J. L., & Verleger, M. A. (2013). The flipped classroom: A survey of the research. Paper presented at the 120th ASEE National conference proceedings, Atlanta, GA.

Bruner, J. S. (1964). The course of cognitive growth. *American Psychologist, 19*, 1-15.

Bruner, J. S., & Kenney, H. J. (1965). Representation and mathematics learning. *The Society for Research in Child Development (monographs), 30*(1), 50-59.

Carroll, J. B. (1964). Words, meanings and concepts. *Harvard Educational Review, 34*(2), 178-202.

Chang, S. H., Lee, N. H., & Koay, P. L. (2017). Teaching and learning with concrete-pictorial-abstract sequence - a proposed model. *The Mathematics Educator, 17*(1), 1-28.

Chao, C. Y., Chen, Y. T., & Chuang, K. Y. (2015). Exploring students' learning attitude and achievement in flipped learning supported computer aided design curriculum: a study in high school engineering education. *Computer Applications in Engineering Education, 23*(4), 514-526.

Chen, L. L. (2016). Impacts of flipped classroom in high school health education. *Journal of Educational Technology Systems, 44*(4), 411-420.

Daher, W., Tabaja-Kidan, A., & Gierdien, F. (2017). Educating Grade 6 students for higher-order thinking and its influence on creativity. *Pythagoras, 38*(1), a350. https://doi.org/10.4102/pythagoras.v38i1.350

Davies, R. S., Dean, D. L., & Ball, N. (2013). Flipping the classroom and instructional technology integration in a college-level information systems spreadsheet course. *Educational Technology Research and Development, 61*(4), 563-580.

DeSantis, J., Van Curen, R., Putsch, J., & Metzger, J. (2015). Do students learn more from a flip? An exploration of the efficacy of flipped and traditional lessons. *Journal of Interactive Learning Research, 26*(1), 39-63.

Engelbrecht, J., Harding, A., & Preez, J. D. (2007). Long-term retention of basic mathematical knowledge and skills with engineering students. *European Journal of Engineering Education, 32*(6), 735-744.

Engin, M. (2014). Extending the flipped classroom model: developing second language writing skills through student-created digital videos. *Journal of the Scholarship of Teaching and Learning, 14*(5), 12-26.

Eustace, B. W. (1969). Learning a complex concept at differing hierarchal levels. Journal of Educational Psychology, *60*(6), 449-452.

Gagné, R. M., & Brigg, L. J. (1979). Principles of instructional design, 2nd ed. New York, NY: Holt, Rinehart and Winston.

Guo, P. J., Kim, J., & Rubin, R. (2014). How video production affects student engagement: an empirical study of MOOC videos. In *Proceedings of the first ACM Conference on Learning@ Scale Conference* (pp. 41-50). New York, NY: ACM.

Handelsman, J., Ebert-May, D., Beichner, R., Bruns, P., Chang, A., DeHaan, R., …Wood, W.B. (2004). Scientific teaching. *Science, 304*(5670), 521-522.

Huang, Y. L. (2016). *Effects of crafted video lessons incorporating multi-modal representations on learning of factorization of quadratic expressions.* (Unpublished master's thesis). National Institute of Education, Nanyang Technological University, Singapore.

Kettle, M. (2013). Flipped physics. *Physics Education, 48*(5), 593-596.

King, F.J., Goodson, L., & Rohani, F. (1998). *Higher order thinking skills: Definitions, strategies, assessment.* Tallahassee, FL: Center for Advancement of Learning and Assessment.

Kwon, O. N., Allen, K., & Rasmussen, C. (2005). Students' retention of mathematical knowledge and skills in differential equations. *School Science and Mathematics, 105*(5), 227-239.

Lai, C. L., & Hwang, G. J. (2016). A self-regulated flipped classroom approach to improving students' learning performance in a mathematics course. *Computers & Education, 100,* 126-140.

Lo, C. K., & Hew, K. F. (2017). A critical review of flipped classroom challenges in K-12 education: possible solutions and recommendations for future research. *Research and Practice in Technology Enhanced Learning, 12*(4), p. 1-22.

Mazur, A. D., Brown, B., & Jacobsen, M. (2015). Learning designs using flipped classroom instruction. *Canadian Journal of Learning and Technology, 41*(2), 1-26.

Medina, J (2014). Brain rules. Seattle, WA, US: Pear Press.

Ng, S. F. (2015). How researchers and teachers could use videos: The role of grounded images. In S. F. Ng (Ed.), *Case of mathematics professional development in East Asian countries* (pp. 1-7). Singapore: Springer.

Nuthall, G. (2000) The role of memory in the acquisition and retention of knowledge in science and social studies units. *Cognition and Instruction, 18*(1), 83-139.

Schultz, D., Duffield, S., Rasmussen, S. C., & Wageman, J. (2014). Effects of the flipped classroom model on student performance for advanced placement high school chemistry students. *Journal of Chemical Education, 91*(9), 1334-1339.

Skemp, R. R. (2006). Relational understanding and instrumental understanding. *Mathematics Teaching in the Middle School, 12*(2), 88-95.

Tsai, C. W., Shen, P. D., & Lu, Y. J. (2015). The effects of problem-based learning with flipped classroom on elementary students' computing skills: a case study of the production of ebooks. *International Journal of Information and Communication Technology Education, 11*(2), 32-40.

Van Eynde, D. F., & Spencer, R. W. (1988). Lecture versus experiential learning: Their differential effects on long-term memory. *Organizational Behavior Teaching Review, 12*, 52-58.

Wang, Y. H. (2016). Could a mobile-assisted learning system support flipped classrooms for classical Chinese learning? *Journal of Computer Assisted Learning, 32*, 391-415.

Wankel, C., & Blessinger, P. (2013). Increasing student engagement and retention in e-learning environments: Web 2.0 and blended learning technologies (Vol. 6). UK: Emerald Group Publishing Limited.

Chapter 10

Designing Mathematical Modelling Activities for the Primary Mathematics Classroom

Chun Ming Eric CHAN   Rashidah VAPUMARICAN
Huanjia Tracy LIU

When students are engaged in mathematical modelling activities, they experience the processes of constructing, explaining, justifying, predicting, conjecturing, quantifying, coordinating, organising, and representing data; processes that are valued in mathematics education. While these are desirable learning outcomes for students to display, conducting such activities in the primary mathematics classroom is rare as the domain of mathematical modelling is especially new in Singapore primary schools and not many teachers know what the modelling activities entail or how to design them. One important aspect is to give consideration to designing such activities towards the realisation of meaningful learning outcomes. This chapter provides a brief perspective to mathematical modelling and addresses the aspect of task design based on two modelling examples to exemplify the alignment of the activities with established design principles.

1 Introduction

Mathematics curricula around the world are recognising the importance of engaging students in rich mathematical tasks or activities with real-world or simulated real-world contexts. Curricula have been reconstructed for future-oriented learning to prepare students to function within a global economy. Recently, numerous countries have embraced STEM

(Science, Technology, Engineering and Mathematics) education in schools (e.g. National Research Council, 2014) with the hope that the nature of such integrated learning experiences will develop students to adapt to real-world problem solving (STEM Taskforce Report, 2014). Moreover, mathematics curricula around the world have increasingly integrated mathematical modelling activities for students to recognise the usefulness of models in today's world as well as develop and use models to interpret and explain structurally complex systems (English, 2011). While there have been interests amongst some teachers in Singapore to carry out mathematical modelling activities (Ang, 2015), support is needed to enhance teachers' knowledge in this relatively new domain. One of the areas is that of designing mathematical modelling activities. This chapter begins with the perspective of mathematical modelling as problem solving and discusses the design of two modelling activities (termed as Model-Eliciting Activities) that have been used in Primary 5 and 6 classes as a case study research. These two sample activities exemplify the considerations involved based on established design modelling principles. A discussion on the practical aspects of designing modelling activities follows.

2 Mathematical Modelling as Real-World Problem Solving

In Singapore, mathematical problem solving is at the heart of the Singapore Mathematics Curriculum Framework (MOE, 2007, 2013). It asserts that the development of five core interrelated components, namely, concepts, skills, attitudes, metacognition and processes are fundamental to developing good problem solvers. It also advocates that the learning of mathematics ought to embrace a dynamic stance comprising the spirit of inquiry and exploration through activity-based learning. One class of activities that encourages the promotion of the dynamic stance of learning is mathematical modelling. The curriculum document states that mathematical modelling is the process of formulating and improving a mathematical model to represent and solve real-world problems (MOE, 2013). Ang (2015) argues that since problem solving is the central theme in Singapore's school mathematics curriculum, it is natural to expect modelling to be part of it in terms of solving and applying mathematics in practical and real-life situations. Chan (2009) contends the need to have a different viewpoint of problem solving, where problem solving entails a number of trial procedures between givens to goals that involve refining and improving one's solutions rather than working through a problem from givens to goals ordered by a set of definitive procedures. The Singapore mathematics curriculum document (MOE, 2013, p.18) informs that during mathematical modelling, students should learn to "deal with ambiguity, make connections, select and apply appropriate mathematics concepts and skills, identify assumptions and reflect on the solutions to real-world problems, and make decisions based on given or collected data" as desirable learning outcomes. In this light, it is not surprising that the notion of problem solving within a modelling context leads to a productive way of thinking as it requires the problem solver to interpret a situation mathematically through iterative cycles of expressing, testing and revising mathematical interpretations of mathematical concepts drawn from various sources (Lesh & Zawojewski, 2007).

3 Model-Eliciting Activities

While there are various perspectives on mathematical modelling in literature, one of the recent modelling perspectives takes the form of engaging learners in model-eliciting activities (MEAs) (Kaiser & Grunewald, 2015). Model-Eliciting Activities are complex, open, non-routine problems situated in real-world contexts that enable learners to exercise both informal and formal mathematics knowledge interactively when engaged (Wessels, 2014). Through sense-making and mathematising, students have to generate models by experiencing the processes of constructing, explaining, justifying, predicting, conjecturing and representing, quantifying, coordinating, organising, and representing data (English, 2003); processes that are valued in mathematics education. A MEA embedded within the context of a real-life situation would make the challenges more personally relevant to students (Kadijevic & Marinkovic, 2006) and also offer challenges at different levels of mathematisation (Freudenthal, 1973). According to Stillman et al. (2009), solving challenging mathematical tasks has the potential of placing the students between their comfort zone and risk taking; challenges teach students to persevere in uncertainty and acquire skills for lifelong learning; and successes with challenges prepare students for real life. English (2003) posits that MEAs provide the vehicle for adequately addressing students' mathematical knowledge, processes, representational fluency, and social skills needed for the 21st century. In MEAs, the models that students produce are conceptual systems made known through the representations that the students have selected to express the idea through written, verbal, or pictorial forms qualitatively or quantitatively (Carmona, 2004).

4 Design Principles

To develop dispositions where students will cultivate a spirit of inquiry and investigation, thoughtful considerations need to be placed on the design of tasks and activities to bring about a more dynamic notion of mathematical activity, after all, "what students learn is largely defined by the tasks they are given" (Hiebert & Wearne, 1997, p.395). In essence, mathematics tasks are usually seen as bridges between students and the learning of mathematics, with the transforming of the subject (mathematics) into activity and tasks through which students can gain access to the mathematics, engage with mathematics, and come to know mathematical concepts (Jaworski, 2014).

We have conducted several case studies on investigating Primary 5 and 6 students' endeavours in mathematical modelling. This involved the designing of modelling activities prior to implementing them. We were guided by the six design principles developed by Lesh, Hoover, Hole, Kelly, and Post (2000) in transforming conceptualisations into MEAs. The principles are:

1. *The Reality Principle*—does the situation resemble some real-life experiences and warrant sense-making and extension of prior knowledge?

2. *The Model Construction Principle*—does the situation create the need to test, modify or extend a mathematically significant construct?

3. *The Self-Assessment Principle*—does the situation require students to self-assess?

4. *The Model Documentation Principle*—does the situation require students to make visible their thinking about the situation?

5. *The Model Shareability Principle*—does the solution serve as a useful model for interpreting other similar situations?

6. *The Simple Prototype Principle*—is the model adequately simple to use?

5 Modelling Examples

Two modelling examples are presented in this chapter. We discuss the use of the six design modelling principles of Lesh at el. (2000) to imply the alignment of the principles used in the design of the modelling activities.

5.1 *Modelling activity 1*

Figure 1 shows a subscription plan modelling task. Students working on this MEA will have to make a recommendation for someone who wishes to subscribe to a phone plan given four different subscription plan options.

1. *Reality Principle*—the context takes the form of a customer wanting to purchase a cellphone and subscribe to a phone plan. As in any real-world contexts today, service providers have a variety of plans to cater to different user needs. The task therefore offers four different subscription plans for students to work on towards determining which would be the best plan to recommend to the customer. The various plans include different rates of calls and free gifts.

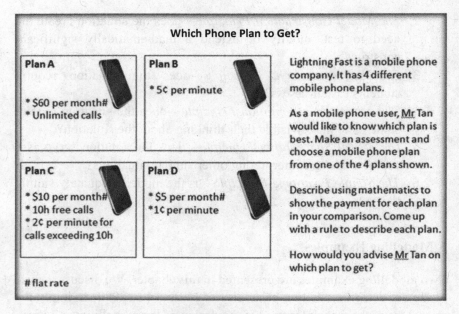

Figure 1. Sample of a phone plan model-eliciting activity

2. *Model Construction Principle*—students can construct a table as a way of modelling to work out the cost of each plan by assuming a fixed time usage for each plan. For example, if the usage is fixed at one hour, what would be the cost for each plan? Students can then fixed another variable, for example, 10 hours and find the cost for each plan. By varying the time usage and thus finding the cost for each plan, they would realise that certain plans may be less expensive than others. Students may also work out a certain mathematical expression, like for Plan B, "5 cents × (number of minutes) or 5 cents × 60 × number of hours" to find the cost of usage. Alternatively, for Plan C, students would have to take into consideration the $10 incurred if the usage is more than 10 hours by crafting "$10 + (2 cents × number of minutes)".

3. *Self-Assessment Principle*—by keeping one variable constant, students have the opportunity to test their conjectures if certain plans would be more viable than others. The various solutions allow for greater discussion leading towards decision making. In this sense, they self-assess

the usefulness of their solutions. For example, at the start, students would tend to think that Plan A is the best because for $60, they have unlimited usage. However, in working out the cost for say 20 hours usage per month, they might find another plan more viable based on consumption needs. So by comparing alternatives, they are able to detect deficiencies and make informed decisions.

4. *Model Documentation Principle*—it is quite impossible for students to work on this modelling task without documenting as they need to constantly make use of the multiple figures in terms of hours used and the cost incurred for each plan in order to make comparisons and decisions. By documenting, we are able to see the assumptions they have made of the variables they keep constant and the solution paths they take towards generating their models. Figure 2 is an example of the solution produced by one group of students. As seen, the students made several assumptions based on hypothetical amounts of time that the customer would use. Based on the parameters given for each plan, the costs were calculated. The documentation is recorded in the form of a very well organized table for ease of comparisons.

5. *Model Shareability Principle*—the models that the students come up with are adaptable to other situations. Besides phone plans, service providers in today's world offer a host of other plans to cater to the needs of the customers, for example, travel plans, cable TV plans, internet plans, and so on. These plans work on the basic idea of cost and usage. By establishing a table for comparing worked data, the models the students produce can become a general way of thinking of a specific solution for a specific context.

6. *Simple Prototype Principle*—the establishing of a table or even a mathematical expression or rule as discussed earlier ensures that the model produced is simple to use. The model can be a learning prototype or metaphor for interpreting other problems with the same underlying structures.

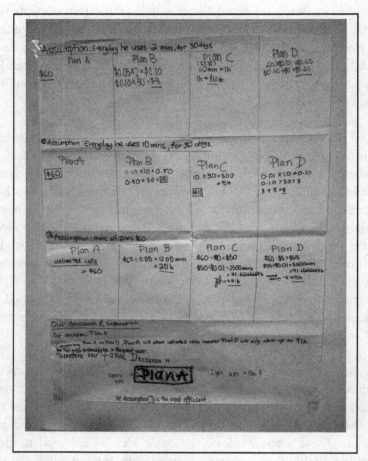

Figure 2. Sample solutions by students of the phone plan MEA

5.2 *Modelling activity 2*

Figure 3 shows a series of slides to enable students to appreciate what happens when clusters of mosquito infection among residents in an island are allowed to grow. The MEA requires students to find out how long would it take a small island to be fully infected given some details about how the number of clusters of infection spreads over time.

1. *Reality Principle*—students are well aware of the national drive to keep mosquito borne diseases down. In schools, they are educated about environmental hygiene as well as the "5-Step Mozzie Wipeout" that can be taken to prevent mosquito breeding. The town councils would also alert their residents should the estates become a cluster area for dengue by placing banners as part of the communication and education drive. Thus, this modelling task taps on the understanding that if citizens do not do their part to prevent mosquito breeding, clusters of infected areas will spread throughout the island. A fictitious name of an island "Zim" is used in this task for simulation purpose to reflect the sense of reality (see Figure 3).

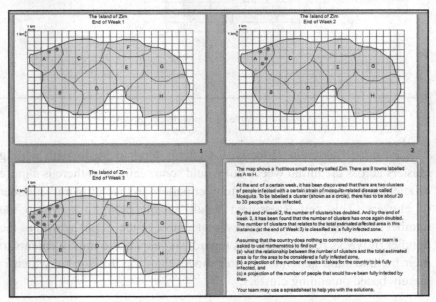

Figure 3. Sample of a mosquito borne disease model-eliciting activity

2. *Model Construction Principle*—students have to ascertain the number of clusters within a certain area and within a certain time (week). They will then look for patterns with respect to the growth of the clusters over several weeks and predict how long it would take the island to be fully infected. Constructing a table aids in presenting the data clearly and helps the students establish the variable relationship or they can even use

a spreadsheet to generate the model. Figure 4 shows an example of a spreadsheet model.

		Week	Clusters	Area	
		1	2		
		2	4		
		3	8	12	
		4	16	24	
		5	32	48	
		6	64	96	
		7	128	192	
		8	256	384	
		9	512	768	
		10	1024	1536	

Figure 4. Sample of a spreadsheet model

3. *Self-Assessment Principle*—students tend to begin with adding clusters across the initial weeks but would soon realise that there is a more efficient way and that is to see a number pattern in the "growing" number of clusters. This self-assessment would enable them to ascertain the number of clusters related to the estimated area of the island as well as the number of weeks required. In this regard, students make improvements to their methods and generate a better model than the initial one, which is merely based on adding clusters.

4. *Model Documentation Principle*—the documenting of students' thinking is necessary as they are working in groups and what is documented becomes the basis for deeper collaborative discussion. The data and figures that they generate often become the reference points for students to ascertain their next course of action. The documentation is more of an audit trail for them to make sense of their own solution paths. Figure 5 is an example of a documented solution by way of their monitoring and assessment of their progress.

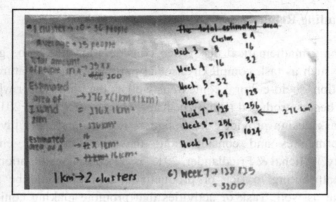

Figure 5. Sample of students' solution

On the left column of Figure 5, students made assumptions using the idea of average in considering the number of people in one cluster. They also estimated the area (i.e. 16 km^2) of part of the island for that part to be considered fully infected. With that area, they worked out the number of weeks for the entire island to be infected as shown in listing on the right column. As the estimated area of the island was found to be 276 km^2, they derived that it would take between 7 to 8 weeks, and found that about 3200 people would be infected by then. The documentation helped them not just in terms of monitoring their thinking, it also made visible their thinking for all members to be aware of how they are progressing towards their goal.

5. *Model Shareability Principle*—the model is a simplistic one and is adaptable for use in other similar simplistic contexts such as finding how soon a certain type of infectious disease may spread or clusters of population may grow based on certain assumptions. If students are able to identify a number pattern or an algebraic expression, these would make the adaptability of the model more convincing.

6. *Simple Prototype Principle*—the use of the spreadsheet suggests that figures may be easily generated and this implies that the model produced is simple and yet transferrable with respect to generating information with similar problem situations.

6 Concluding Remarks

In designing mathematical tasks, due consideration has to be given to elements such as task complexity, social participation, the nature of the participation, socio-cultural context, as well as the purpose for which the task has been introduced (Shimizu, Kaur, Huang, & Clarke, 2010). The role of task design provides the designers to "explicate and investigate design principles and considerations that are frequently employed intuitively" (Michal & Friedlander, 2012, p. 33). The consideration of task features attributing to rich or productive learning should be seen as significant as well. Task or activities that promote making conjectures, abstracting mathematical properties, reasoning, validating, discussing, and question eliciting ought to be embedded in task features for students to develop such skills as well as exhibit them. Furthermore, situating the task within the context of a real-life problem provides a more "authentic" aspect to solving problems such as managing data in less-structured or ill-structured problem situations. We contend that the use of MEAs serves as an excellent platform for students to develop the repertoire of processes valued in mathematics education. Lesh et al. (2000) have promoted six modelling design principles: reality, model construction, self-assessment, documentation, shareability and simple prototype principles. Designs should preferably adhere to the said principles to facilitate extended learning for the students but we are also mindful that we are creating learning opportunities for young students and there could be times when certain principles are not satisfied and some circumvention of the principles might be necessary.

For teachers who wish to design and carry out MEAs in the class, there is a need to get acquainted with what MEAs are first. Chamberlin and Moon (2005) assert it that while it is possible for anyone to create MEAs, it is wise to at least receive minimal training from experts in this area. As well, it will be worthwhile to find out more through various channels before implementing MEAs. The internet provides a rich host of websites carrying this subject and one would notice that task demands and complexities vary for MEAs. Books written by experts and practitioners on this subject will also provide interested learners with a generally good grounding. This chapter comes about as a result of a collaboration between teachers and a researcher who are interested to carry out MEAs in the primary mathematics classroom.

In designing a MEA, consider what might be some interesting mathematical relationships to explore and at the same time assess if they are within the capabilities of the students to work on. Consider situations that students see around them in and beyond the classroom that may be viable for use as a problem context. For a start, one may design a simple modelling activity such as having students explore shoe-size and height relationships towards expressing that relationship. Next carry it out with the class to get a feel of the facilitation of the activity as well as observe what the students are discussing about. According to Ang (2015), such activities may be classified as Level 1 type as the focus may be on acquiring skills that are directly or indirectly related to mathematical modelling such as drawing a graph to depict a relationship. This start-small experience may be helpful towards scaling up the design to slightly more challenging modelling activities such as the examples presented earlier in this chapter. The six design principles should be factored in, as these will provide the guide in ensuring students are producing some expected and desirable outcomes when they are engaged with MEAs. More complex modelling activities should aim to develop higher level modelling competencies such as the ability to apply specific knowledge such as making assumptions or making interpretations in mathematical modelling (Level 2) or at the highest level, to work collaboratively, carry out discussions, develop a model, solve the model and make presentations (Level 3). With the modelling design principles as a blueprint, designing modelling activities can be a meaningful endeavour in creating rich mathematical tasks and activities.

References

Ang, K. C. (2015). Mathematical Modelling in Singapore Schools: A Framework for Instruction. In N. H. Lee, & K. E. D. Ng (Ed), *Mathematical modelling - from theory to practice* (pp 57-72), World Scientific: Singapore.

Chamberlin, A., & Moon, M. (2005). Model eliciting activities as a tool to develop and identify creatively gifted mathematicians. *The Journal of Secondary Gifted Education, 17*(1), 37-47.

Carmona, G. (2004). *Designing an assessment tool to describe students' mathematical knowledge*. Purdue University, West Lafayette, IN.

Chan, C. M. E. (2009). Mathematical modelling as problem solving for children in the Singapore Mathematics Classroom. *Journal of Science and Mathematics Education in Southeast Asia, 32*(1), 36-61.

English, L. D. (2011). Complex Learning through cognitively demanding tasks. *The Mathematics Enthusiast, 8*(3), 483-506.

English, L. D. (2003). Mathematical modelling with young learners. In S. J. Lamon, W. A. Parker, & K. Houston (Eds). *Mathematical modelling: A way of life – ICTMA 11* (pp. 3-18). Horwood Publishing Limited.

Freudenthal, H. (1973). *Mathematics as an educational task*. Dordrecht, The Netherlands: Reidel.

Hiebert, J., & Wearne, D. (1997). Instructional tasks, classroom discourse and student learning in second grade arithmetic. *American Educational Research Journal, 30*, 393-425.

Jaworski, B. (2014). Mathematics education development. Research in teaching ↔ Learning in practice. In J. Anderson, M. Cavanagh, & A. Prescott (Eds.), Curriculum in focus: Research guided practice. *Proceedings of the 37th Annual Conference of the Mathematics Education Research Group of Australasia*, (pp. 2-23). Sydney: MERGA.

Kaiser, G., & Grunewald, S. (2015). Promotion of mathematical modelling competencies in the context of modelling projects. In N. H. Lee, & K. E. D. Ng (Eds.), *Mathematical modelling: from theory to practice* (pp. 21-40). Singapore: World Scientific.

Kadijevic, D., & Marinkovic, B. (2006). Challenging mathematics by "Archimedes". *The Teaching of Mathematics, 9*(1), 31-39.

Lesh, R., Hoover, M., Hole, B., Kelly, A., & Post, T. (2000). Principles for developing thought-revealing activities for students and teachers. In A. Kelly, & R. Lesh (Eds.), *Handbook of research design in mathematics and science education* (pp. 591-645). Mahwah, NJ: Lawrence Erlbaum.

Lesh, R., & Zawojewski, J. (2007). Problem solving and modeling. In F. K. Lester (Ed.), *Second handbook of research on mathematics teaching and learning: A project of the National Council of Teachers of Mathematics* (pp. 763-803). Charlotte, NC: Image Age Publishing.

Michal, T., & Friedlander, A. (2012). Five considerations in task design: The case of improving grades. *Investigations in Mathematics Learning, 4*(3), 32-49.

MOE. (2007). *Ministry of Education Mathematics Syllabus* – Primary, Singapore: Curriculum Planning and Development Division.

MOE. (2013). *Ministry of Education Mathematics Syllabus* – Primary, Singapore: Curriculum Planning and Development Division.

National Research Council. (2014). *STEM learning is everywhere: summary of a convocation on building learning systems*. Washington, DC: The National Academies Press.

Shimizu, Y., Kaur, B., Huang, R., & Clarke, D. (2010). The role of mathematical tasks in different cultures. In Y. Shimizu, B. Kaur, R. Huang, & D. Clarke (Eds.), *Mathematical tasks in classrooms around the world* (pp.1-14). Rotterdam, The Netherlands: Sense Publishers.

Stillman, G., Cheung, K., Mason, R., Sheffield, L., Sriraman, B., & Ueno, K. (2009). Challenging Mathematics: Classroom Practices. In E. J. Barbeau, & P. J. Taylor (Eds.), *Challenging Mathematics In and Beyond the Classroom. The 16th ICMI Study* (pp. 243-284). New York: Springer.

Task Force Report, STEM. (2014). Innovate: a blueprint for science, technology, engineering, and mathematics in California public education. Dublin, California: Californians Dedicated to Education Foundation. http://www.cde.ca.gov/nr/ne/yr14/yr14rel71.asp

Wessels, H. (2014). Levels of mathematical creativity in model-eliciting activities. *Journal of Mathematical Modelling and Application, 1*(9), 22-40.

Chapter 11

Extending Textbook Exercises into Short Open-Ended Tasks for Primary Mathematics Classroom Instruction

YEO Kai Kow Joseph

Textbook exercises do not always lend themselves to multiple solutions and reasoning. Many problems can be extended to open-ended tasks and made accessible to pupils with varying abilities. Instead of creating their own open-ended tasks, a more feasible approach is for teachers to learn how to convert routine textbook exercises into short open-ended tasks. The main aim of this chapter is to focus on the types of short open-ended tasks that primary mathematics teachers can extend from closed problems found in textbooks. This chapter therefore reviews the mathematical tasks used in classroom instruction and the benefits of implementing open-ended tasks. This chapter also discusses the implication of teaching using open-ended tasks at the primary level.

1 Introduction

For more than two decades, the role of textbooks in mathematics teaching and learning has received increasing attention from the international mathematics education arena. For instance, in the Third International Mathematics and Science Study (TIMSS) organised by the International Studies in Educational Achievement (IEA), textbooks were established as an independent study for the first time ever in a research of this scope (Schmidt, McKnight, Valverde, Houang, & Wiley, 1997). Moreover, textbooks do not just focus on intentions and aims of the curricula (Schmidt et al., 1997) but also have a great influence on mathematics instruction

(Howson, 1995). Problems have always been considered an essential part of a mathematics lesson to introduce a new topic or to assess understanding. Zhu (2003) investigated how different types of problems are represented in three countries' mathematics textbooks and reported that almost all the problems in the examined books provide just sufficient information. It seems that the provision of too many problems with just sufficient information in mathematics textbooks could be a possible reason pupils often failed to distinguish extraneous information from the necessary, and tended to use all the information provided in the problems (Carpenter, Corbitt, Kepner, Lindquist, & Reys, 1980).

The framework of the Singapore Curriculum that was conceptualised during the early 1990s to encompass mathematical problem solving as its central focus, advocated that problems should cover a wide range of situations from routine mathematical problems to problems in unfamiliar contexts and open-ended investigations that make use of relevant mathematics and thinking processes (Ministry of Education, 2000). It is evident from the framework that problems are both a means and an end in Singapore school mathematics instruction. In the context of Singapore, the typical primary and secondary school textbook exercises, worksheets and assessment test items are frequently presented in closed problem formats. They are created with a narrow focus on a few specific skills, concepts and procedures that have been previously taught. They constitute the "mechanical" part of learning and practicing mathematical procedures and formulas. There are also non-routine word problems in the textbooks that are presented in mathematics lessons where the emphasis is on the practice of heuristics on similar kinds of problems. However, textbook exercises do not always lend themselves to multiple solutions and reasoning. Many problems can be extended to open-ended tasks, and made accessible to pupils with a wide range of abilities with minimum effort. Instead of creating their own open-ended tasks, a more feasible approach is for teachers to learn how to convert routine textbook exercises into short open-ended tasks. The main aim of this chapter is to focus on the types of short open-ended tasks that mathematics teachers can extend from closed problems found in textbooks. This chapter therefore reviews the mathematical tasks used in classroom instruction and the benefits of implementing open-ended tasks. This chapter also discusses the implication of teaching using open-ended tasks at the primary level.

2 Review of Literature

This section explains the role of mathematical tasks used in classroom instruction. In addition, reviews from various researchers and educators on the benefits of implementing open-ended tasks in the classroom are also discussed.

2.1 *Mathematical tasks used in classroom instruction*

The National Council of Teachers of Mathematics (2000) envisaged mathematical tasks to be important to the learning of essential mathematics. Mathematical tasks offer a key prompt for enabling learning and communication in the mathematics lesson (Stein & Lane, 1996; Sullivan, Warren, & White, 2000; Viseu & Oliveira, 2012). Krainer (1993) recognized, "Powerful tasks are important points of contact between the actions of the teacher and those of the student" (p. 68). Lappan and Briars (1995) also emphasized that choosing tasks or problems is a major decision that could affect pupils' learning. As mathematical tasks play such a significant part in the effectiveness of mathematics teaching, how do teachers choose appropriate tasks to enhance pupils' learning? One way encompasses the choice of "high-level cognitively complex tasks [that can promote] the capacity to think, reason and problem solve" (Smith & Stein, 1998, p.344). This emphasis is established on the notion that if pupils are challenged at an appropriate level with non-routine tasks, they expand their cognitive abilities and engage in rich mathematical discussions. Undeniably if more time were spent in classrooms with pupils engaged in solving cognitively demanding non-routine tasks, as opposed to exercises in which a known algorithm is practiced, pupils' opportunities for thinking and learning would likely be improved. Another common way is to choose mathematical tasks that activate engagement with the mathematical concept to be learned (Bell, 1993; van Boxtel, van der Linden, & Kanselaar, 2000). This process can also provide opportunities for student learning.

Mathematics teachers can modify the characteristics of the problem, depending on the specific instructional objective, the intended mathematical knowledge and the range of pupils' abilities (Sullivan & Clarke, 1992). Open-ended tasks, in particular, have been considered effective for creating opportunities for student exploration, collaboration, and lucid mathematical reasoning (Kosyvas, 2016). However, Clarke (2011) contended that open-ended tasks are difficult to define due to the dependence of task interpretation on the social condition in which the task is outlined and undertaken. Although there are various views of open-ended tasks, we can still identify some common characteristics. Open-ended tasks "have more than one answer and/or can be solved in a variety of ways" (Moon & Schulman, 1995, p. 25). Hancock (1995) also agreed that open-ended tasks could be considered to have more than a single correct solution and that they offer pupils multiple approaches to the tasks by placing little constraints on the pupils' methods of solution.

2.2 *Benefits of implementing open-ended tasks*

There continue to be many benefits in implementing open-ended tasks in mathematics instruction. Although there is no all-encompassing definition of an open-ended task, we can still establish some common characteristics of how the types of open-ended tasks can be used to motivate pupils' learning of mathematics. London (1993) for instance defined some traits of open-ended tasks. London indicated that open-ended tasks entail the three phases of task recognition, trial, and perseverance. They should have different solutions and they can be used to assess pupils' learning. Finally, they should be solvable by any pupil and pupils would take some time to solve the problem. Sawada (1997) further listed five benefits of open-ended tasks. First, pupils take part in class more enthusiastically and express their thoughts more freely. Second, pupils can have a greater opportunity to use their mathematical ideas and skills more widely. Third, all pupils can answer the problem in their own meaningful ways. Fourth, mathematics lessons with open-ended tasks provide pupils with enriching learning experiences. Fifth, pupils experience the fulfillment of discovery and the approval of other pupils. Finally, Yeo (2016) proposed that open-ended tasks are those that require pupils to think more intensely and to provide a solution that involves more than remembering a fact or repeating a skill. There are many possible solution approaches and strategies to solve an open-ended task. The focus of open-ended tasks offers opportunities for pupils to reveal their decision making process, mathematical thinking, reasoning process as well as problem solving and communication skills.

There are also other benefits in implementing open-ended tasks. For instance, specific studies that supported the use of open-ended tasks include Stein and Lane (1996) who noted that student performance gains were greater when "tasks were both set up and implemented to encourage use of multiple solution strategies, multiple representation and explanations" (p. 50). Furthermore, when a pupil learns mathematics through such an open-ended approach, struggling with the difficulties facing him instead of relying on memorization or any pre-determined rules to search for solutions, it promotes "deep understanding" of the mathematics (Hiebert et al., 1996). In fact, using the open approach to teaching mathematics can potentially change both pupils' and teachers' attitudes and beliefs about mathematics and develop their reasoning, communication skills, and making connections in mathematics (Hashimoto & Becker, 1999). Given that open-ended tasks are selected for all pupils, regardless of their mathematical abilities, skills, and interests, they are not intended only for pupils with high motivation and skills. Nohda (1986) indicated that even pupils with a lower motivation level could be involved in solving open-ended tasks. Appealing to pupils' natural ways of thinking lies at the heart of the "open approach," as pupils will produce different solutions and then will share their thinking with the rest of the class.

At the heart of it all, open-ended tasks can give pupils a sense of achievement and fulfillment because it is possible even for low progress learners with less mathematical ability to write their own solutions within their own ability. Furthermore, it offers pupils a chance to feel what it is like to be real mathematics learners in the course of solving open-ended tasks. Here, what is noteworthy is that both teachers and pupils identify the solver's contribution in the learning process and that every solver has confidence in his or her ability to find his or her own solutions.

3 Extending Textbook Exercises to Short Open-Ended Tasks

The brief review of literature in the previous section has provided an outline of the benefits of open-ended tasks. One cannot count on the standard mathematical textbook exercises used by teachers to support this new view of mathematics education. Pupils must work on rich mathematical tasks that provide opportunities for them to reason, offer evidence for their thinking, communicate and present their ideas, as well as find connections across mathematical ideas. With the introduction of the 2013 Mathematics syllabus in Singapore, it is appropriate to revisit open-ended tasks in terms of their relevance and applicability to the teaching of primary mathematics. It is a challenge to create different types of open-ended tasks. For this reason, the open-ended tasks with the following two features are discussed in this chapter. First, the initial point of the task is rather clear but the solutions for its objectives can vary. Second, they are tasks in which pupils can show higher-order thinking skills and employ divergent thinking in the search of their own solutions. This section focuses on short open-ended tasks that teachers can extend from closed problems found in textbook and workbook exercises. Teachers can also use these short open-ended tasks for when "teaching through problem solving". The following subsections exemplify open-ended tasks that make assumptions on the missing information, and open-ended tasks that discuss a concept, algorithm or error (Yeo, 2016), as two possible categories that can be used in the mathematics classroom instruction at the primary level.

3.1 *Open-ended tasks that make assumptions on the missing information*

This type of open-ended task requires pupils' own contribution to the process such as making assumptions on the missing information and there is no fixed process that guarantees a correct answer. In addition, pupils also need to solve the problem where there is no known solution beforehand and not all data are given. The content specific open-ended tasks in this category can be illustrated by means of two examples below.

Consider, for example, a typical textbook problem on the difference of two numbers (see Figure 1). Textbook Problem 1 has a right or wrong answer and is solved in one-step and one way. Normally, a pupil will work individually to find the answer after applying the subtraction algorithm. Such problems enable the teacher to know whether the pupil has learned the procedure for finding the difference of two numbers. However, Textbook Problem 1 does not enhance a pupil's understanding on the difference of two numbers. To make it more open-ended, I have extended Textbook Problem 1 on finding the difference of two-digit numbers into an open-ended situation for pupils to choose a pair of numbers and compare the two numbers so that the pupil can subtract the smaller number from the bigger number (see Figure 1). Not only are pupils using thinking skills, they are also getting valuable practice subtracting multiple combinations of numbers. The responses received from pupils to the open-ended task give the teacher a deeper insight into the pupils' real understanding of comparing numbers and subtraction of two numbers. For instance, a primary 2 pupil's response to Open-Ended Task 1, shown in Figure 2, would have been scored as correct in the closed problem. However, the written verbal response by the primary 2 pupil, shown in Figure 2, was: "I get the pairs difference by the greater number taken away by the smaller number". This showed that the pupil would need remedial assistance to use mathematical language to express mathematical ideas precisely, concisely and logically. The pupil's workings showed how he had made some decision to choose and compare two different numbers and had the correct understanding of the meaning of difference of two numbers.

Textbook Problem 1:
What is the difference between 90 and 12?

Open-Ended Task 1:
Choose a pair of numbers. Then find the difference between the two numbers. Now challenge yourself by choosing many pairs of numbers and find their difference. Explain clearly how you find the difference for each pair.

Figure 1. Extending calculations problems to Open-Ended Task 1

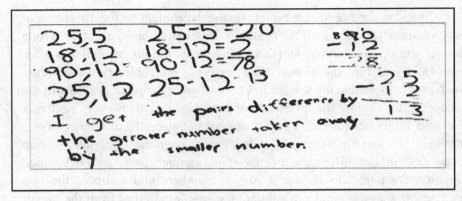

Figure 2. A primary 2 pupil's response to Open-Ended Task 1

The next example in Figure 3 shows a typical primary one textbook problem on finding the number of sides and corners for different shapes. Textbook Problem 2 again has a right or wrong answer and pupils can just identify and indicate the number of sides and corners for each shape. Usually, a pupil will identify and count the number of sides and corners of a shape and then fill in the blanks. However, Textbook Problem 2 does not allow the pupil to make a decision to choose and describe any 2-dimensional shape according to attributes such as sides, corners, sizes, colours and orientation. To make it more open-ended, I have extended this textbook problem on finding the number of sides and corners for a shape into an open-ended situation where primary 1 pupils need to visualize and draw their favourite shape as well as write the number of sides and corners. For instance, a primary one pupil's response to Open-Ended Task 2, shown in Figure 4, had made a decision to choose a triangle as his favourite shape and he was able to write the correct number of sides and corners. He also drew many different sizes of triangles in different orientation.

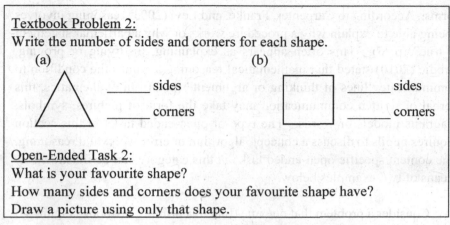

Figure 3. Extending a geometry problem to Open-Ended Task 2

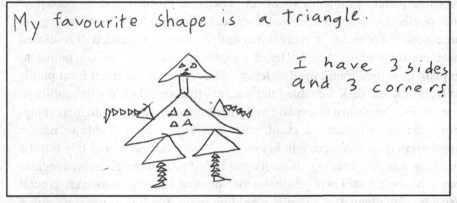

Figure 4. A primary 1 pupil's response to Open-Ended Task 2

3.2 *Open-ended tasks to discuss a concept, algorithm or error*

Reasoning is a process that underpins mathematical thinking and is one in which the mathematics curriculum in Singapore would expect teachers to facilitate in classroom instruction (Ministry of Education, 2012). While past syllabus documents had recommended that mathematical reasoning be taught and learnt in classrooms, its enactment is often unclear and seldom emphasised. Moreover, mathematical reasoning has been defined in various forms. According to Carpenter, Franke, and Levi (2003) reasoning involves "being able to explain why a procedure works or why a particular statement is true" (p. 5). This often appears as explaining, justifying or proving. Brodie (2010) stated that mathematical reasoning assumes the condition to communicate "lines of thinking or argument" (p. 7). In mathematics, this verbal or written communication may take the form of pictures, symbols, diagrams models or words. The type of open-ended tasks in this section requires pupils to discuss a concept, algorithm or error with valid reasoning. The content specific open-ended tasks in this category can be illustrated by means of two examples below.

Consider a problem that one may find in a primary textbook (Figure 5). Formulated in such a closed structure, the teacher and pupils would have in mind the expected standard response is to find the correct 36th letter in this repeated pattern problem. However, Textbook Problem 3 does not allow the pupils to analyse the mathematical situation and provide logical argument. To make it more open-ended, I have extended this closed problem on finding the 36th letter into an open-ended task for pupils to explain how they found the 36th letter. The responses received from pupils to Open-Ended Task 3 enabled the teachers to assess their pupils' ability to reason and the ability to explain and justify their answers or their working. For instance, a primary 2 pupil, shown in Figure 6, was able to reason appropriately and correctly in his written verbal response, and this pupil's thinking was enlightening. If such open-ended tasks were given on a regular basis, it would instill in pupils that understanding and explanation are crucial aspects of mathematics. Pupils would be more mindful to use the correct mathematical language to express mathematical ideas precisely and concisely.

The next example in Figure 7 shows a typical primary 2 textbook problem on division. Textbook Problem 4 again has a right or wrong answer and pupils may just recall the multiplication table of three to fill in the answer. To make it more open-ended, I have extended this division sum into an open-ended situation where pupils need to explain how they could find the value of $27 \div 3$ from $18 \div 3$. It is conceivable that within the same class, some pupils may offer responses by drawing 27 items and putting them in groups of three and thus offering evidence of their understanding of division as separating into equal groups. However, within the same class, some pupils may be also operating at a different level. For instance, a primary 2 pupil (see Figure 8) was able to begin from $18 \div 3$ and add three "3s" to show $27 \div 3 = 9$. Such responses may not have emerged when using textbook closed problems that have predetermined methods and answers. By modifying the problems in this way, teachers are able to gain access to more knowledge of their pupils' levels of understanding. This approach enabled teachers to see the pupil's thinking rather than their own thinking.

Textbook Problem 3:

A B A B A B A B A B A B A B A B A B A ...

This pattern of letters continues.

What is the 36th letter in the line?

Open-Ended Task 3:

A B A B A B A B A B A B A B A B A B A ...

This pattern of letters continues.

Your classmate asks you, "What is the 36th letter in the line?"

Explain how you found the letter.

Figure 5. Extending a pattern problem to Open-Ended Task 3

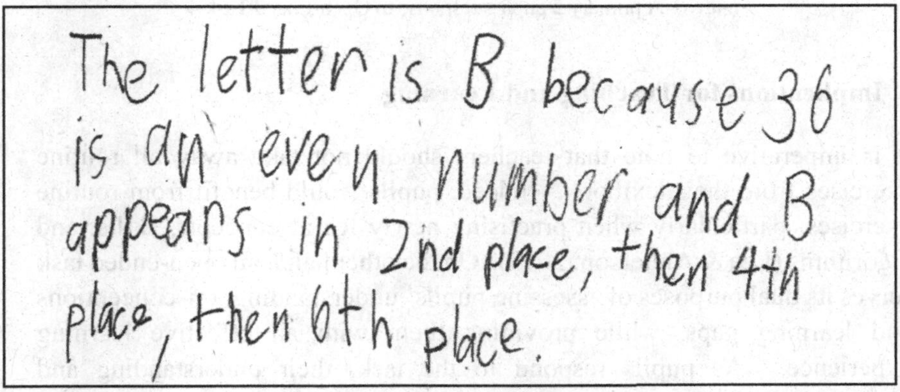

Figure 6. A primary 2 pupil's response to Open-Ended Task 3

Textbook Problem 4:
Fill in the missing number in the boxes.

$18 \div 3 =$ ☐ $27 \div 3 =$ ☐

Open-Ended Task 4:
Jane has forgotten what is $27 \div 3$, but she remembers that $18 \div 3 =$ is 6. Can you help Jane?
Explain clearly the strategy that you use.

Figure 7. Extending a division sum to Open-Ended Task 4

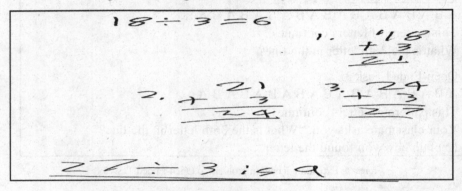

Figure 8. A primary 2 pupil's response to Open-Ended Task 4

4 Implications for Teaching and Learning

It is imperative to note that teachers should not take away all routine exercises from their textbook. Indeed, pupils would benefit from routine exercises, particularly when practising newly learnt concepts, skills, and algorithms (Lee & Anderson, 2013). On the other hand, an open-ended task serves its dual purposes of assessing pupils' understanding, misconceptions and learning gaps, while providing them with an effective learning experience. As pupils respond to the task, their understanding and connections of mathematical concepts extend. Through these short open-ended tasks, teachers can assess much more than whether or not pupils can solve problems or perform straightforward computations. Pupils' conceptual understanding of various embedded knowledge can be assessed and teachers can use that information to plan and improve their teaching and learning.

Task design can form a critical part of lesson planning, particularly for mathematics teachers (Choy, 2016; Roche, Clarke, Clarke, & Sullivan, 2014). Mathematics teachers should also be mindful when they design open-ended tasks for different learners. There is a delicate balance between allowing pupils to persevere through the open-ended task and providing them support before they become too frustrated and give up. Mathematics teachers must know their pupils well, modify the task or the time allotted for the task appropriately for individual pupils and the class, and know when to provide assistance. For instance, low progress learners with greater gaps in understanding tend to learn much from open-ended tasks, although at a slower pace than others. At the initial stage, low progress learners may pause at solving these open-ended tasks, as they are not given explicit instructions on how to complete the task or what the correct response may be. However, as teachers introduce these open-ended tasks more regularly, low progress learners will be more ready to solve them. For these learners, it may be advisable to scaffold their experiences at first by using versions of tasks that are differentiated for their specific needs before using more stimulating open-ended tasks. As these low progress learners are more prone to give up easily in problem solving and may have difficulty persevering in such open-ended tasks, care must be taken not to discourage them. Consequently, it is prudent to limit the duration of the tasks at the outset and increase the duration of tasks progressively. Allowing pupils to solve open-ended tasks as a pair, rather than individually, may also support them and reduce the chance of them being overly discouraged. In contrast, high progress learners who are competent with more advanced mathematics should also be given open-ended tasks that also meet their needs. Most open-ended tasks can be easily made deeper and more thought provoking. These high progress learners often enjoy such open-ended tasks. High progress learners can be given these open-ended tasks prior to instruction on particular topics. They can learn through these tasks, sometimes even independently of the mathematics teacher. Furthermore, high progress learners can also be asked to pose and solve their own open-ended tasks, leading to challenging learning experiences.

Open-ended tasks may seem more mysterious initially than textbook exercises, particularly if they are unusual or unfamiliar. Although pupils may hesitate to solve these open-ended tasks at first, class time must be planned for pupils to struggle with and learn through an open-ended task. Teachers must place some trust in pupils as learners and communicate high expectations to them. When pupils show major learning gaps, teachers can provide just-in-time instructions directed at particular concepts and scaffold understanding. For mathematics teachers who have not begun on such an open-ended approach, it is time to try.

5 Concluding Remarks

Teachers interested in supporting the development of pupils' flexible and creative thinking need to be able to differentiate between textbook routine exercises and open-ended tasks and, when appropriate, extend textbook routine exercises into short open-ended tasks. Instead of continuing to give the same amount of textbook exercises or worksheets, reduce some textbook routine problems and make way for pupils to solve some open-ended tasks in class. Using open-ended tasks in the classroom can open a new window into our pupils' mathematical understanding. Specifically, if we expect pupils to show their conceptual and relational understanding of mathematics, teachers must give them the opportunities to reveal that understanding. In addition, mathematics teachers must shift paradigms towards a more process-based approach in which getting a correct answer to a problem is not the main measure. In implementing open-ended tasks, teachers must recognise the embedded mathematical knowledge and connections that might develop. In order not to impede pupils' ingenuity in mathematical thinking, teachers must also proactively and consistently support pupils' cognitive activity without reducing the cognitive demands of the open-ended tasks. When pupils make their mathematical thinking visible through reasoning, explanations and multiple responses, teachers gain insight into the robustness of their understanding. Those insights can then be used to fine-tune the teachers' teaching so that learning is enriched.

References

Bell, A. (1993). Principles for the design of teaching. *Educational Studies in Mathematics,* *24,* 5-34.

Brodie, K. (2010). *Teaching mathematical reasoning in secondary school classrooms.* New York: Springer.

Carpenter, T. P., Corbitt, M. K., Kepner, H. S., Jr., Lindquist, M. M., & Reys, R. E. (1980). Solving verbal problems: Results and implications from national assessment. *Arithmetic Teacher, 28*(1), 8-12.

Carpenter, T. P., Franke, M., & Levi, L. (2003). *Thinking mathematically: Integrating arithmetic and algebra in elementary school.* Portsmouth: Heineman.

Choy, B. (2016). Snapshots of mathematics teacher noticing during task design. *Mathematics Education Research Journal, 28*(3), 421-440. doi:10.1007/s13394-016-0173-3

Clarke, D. J. (2011). Open-ended tasks and assessment: The nettle or the rose. In B. Kaur, & K. Y. Wong (Eds.), *Assessment in the mathematics classroom* (pp. 131-163). Singapore: World Scientific.

Hancock, C.L. (1995). Enhancing mathematics learning with open-ended questions. The *Mathematics Teacher, 88*(6), 496-499.

Hashimoto, Y., & Becker, J. P. (1999). The open approach to teaching mathematics-Creating a culture of mathematics in the classroom: Japan. In L. Sheffield (Ed.), *Developing mathematically promising students.* Reston, VA: National Council of Teachers of Mathematics.

Hiebert, J., Carpenter, T. P., Fennema, E., Fuson, K., Human, P., Murray, H., ... Wearne, D. (1996). Problem solving as a basis for reform in curriculum and instruction: The case of mathematics. *Educational Researcher, 25*(4), 12-21.

Howson, A. G. (1995). *TIMSS monograph no. 3: Mathematics textbooks: A comparative study of grade 8 texts.* Vancouver, BC: Pacific Educational Press.

Kosyvas, G. (2016). Levels of arithmetic reasoning in solving an open-ended problem. *International Journal of Mathematical Education in Science and Technology, 47*(3), 356-372. doi:10.1080/00207 39x.2015.1072880

Krainer, K. (1993). Powerful tasks: A contribution to a high level of acting and reflecting in mathematics instruction. *Educational Studies in Mathematics, 24,* 65-93.

Lappan, G., & Briars, D. (1995). How should mathematics be taught? In I. M. & Carl (Eds.), *Seventy-five years of progress: Prospects for school mathematics* (pp. 115-156). Reston, VA: National Council of Teachers of Mathematics.

Lee, H. S., & Anderson, J. R. (2013). Student learning: What has instruction got to do with it? *Annual Review of Psychology, 64,* 445-469.

London, R. (1993). *A curriculum of nonroutine problems.* Paper represented at the Annual Meeting of the American Educational Research Association, Atlanta, GA, (ERIC Document Reproduction Service No. ED359213).

Ministry of Education (2000). *Mathematics syllabus: Primary.* Singapore: Curriculum Planning and Development Division.

Ministry of Education (2012). *Primary mathematics: Teaching and learning syllabus.* Singapore: Curriculum Planning and Development Division.

Moon, J., & Schulman, L. (1995). *Finding the connections: Linking assessment, instruction, and curriculum in elementary mathematics.* Portsmouth, N.H: Heinemann.

National Council of Teachers of Mathematics. (2000). *Principles and standards for school mathematics.* Reston, VA: Author

Nohda, N. (1986). A study of "open-approach" method in school mathematics. *Tsukuba Journal of Educational Study in Mathematics, 5,* 119-131.

Roche, A., Clarke, D. M., Clarke, D. J., & Sullivan, P. (2014). Primary teachers' written unit plans in mathematics and their perceptions of essential elements of these. *Mathematics Education Research Journal, 26*(4), 853-870. doi:10.1007/s13394-014-0130-y

Sawada, T. (1997). Developing Lesson Plans. In J. Becker, & S. Shimada (Eds.), *The open-ended approach: A new proposal for teaching mathematics.* (p. 23-35). National Council of Teachers of Mathematics.

Schmidt, W. H., McKnight, C. C., Valverde, G. A., Houang, R. T., & Wiley, D. E. (1997). *Many visions, many aims: A cross-national investigation of curricular intentions in school mathematics.* Boston: Kluwer Academic Press.

Smith, M. S., & Stein, M. K. (1998). Selecting and creating mathematical tasks: From research to practice. *Mathematics Teaching in the Middle School, 3,* 344-350.

Stein, M. K., & Lane, S. (1996). Instructional tasks and the development of student capacity to think and reason and analysis of the relationship between teaching and learning in a reform mathematics project. *Educational Research and Evaluation, 2*(1), 50-80.

Sullivan, P., & Clarke, D. J. (1992). Problem solving with conventional mathematics content: Responses of pupils to open mathematical tasks. *Mathematics Education Research Journal, 4*(1), 42-60. doi:10.1007/bf03217231

Sullivan, P., Warren, E., & White, P. (2000). Students' responses to open ended mathematical tasks. *Mathematics Education Research Journal, 21*(1), 2-17.

van Boxtel, C., van der Linden, J., & Kanselaar, G. (2000). Collaborative learning tasks and the elaboration of conceptual knowledge. *Learning and Instruction, 10,* 311-330.

Viseu, F., & Oliveira, I. B. (2012). Open-ended tasks in the promotion of classroom communication in mathematics. *International Electronic Journal of Elementary Education, 4*(2), 287-300.

Yeo, K. K. J. (2016). Using Open-Ended Tasks to Foster 21st Century Learners at the Primary Level. In P. C. Toh, & B. Kaur (Eds.), *Developing 21st Century Competencies in the Mathematics Classroom* (pp. 132-148). Singapore: World Scientific.

Zhu, Y. (2003). *Representations of problem solving in China, Singapore and US mathematics textbooks: a comparative study.* Unpublished doctoral dissertation. National Institute of Education, Singapore.

Chapter 12

Integrating Problem Posing into Mathematical Problem Solving: An Experimental Study

JIANG Chunlian   CHUA Boon Liang

Mathematical problem posing (PP) has been emphasized in mathematics curricula around the world. Integrating PP into mathematics instruction has grown steadily among mathematics education researchers and practitioners. Quite a few studies used the what-if-not strategy proposed by Brown and Walter (1990) to help students to develop their PP skills. In a project on mathematical situations and PP, Lv and Wang (2006) proposed a teaching model to integrate PP into mathematics instruction. In this paper, we demonstrate how a mathematical problem for first graders was used to carry out the what-if-not strategy to pose problems first, then describe a lesson that follows strictly the instruction model developed by Lv and Wang (2006), and finally report the results obtained from an experimental study with 56 seventh grade Macao students. The results indicated that the experimental design with PP integrated did improve students' PP skills.

1 Introduction

In recent years, interest in incorporating mathematical problem posing (PP) in school mathematics instruction has gained greater interest and attention among mathematics education researchers and practitioners (e.g. Cai, Hwang, Jiang, & Silber, 2015; Singer, Ellerton, & Cai, 2013, 2015). Curriculum reforms in many countries, including Singapore, have begun to raise the profile of PP at different educational levels over the past several decades (e.g., van den Brink, 1987; Brown & Walter, 1993; Chinese Ministry of Education, 1986, 1992, 2000a, 2000b, 2001, 2011; Hashimoto, 1987; Healy, 1993; Keil, 1967; Kilpatrick, 1987; NCTM, 2000; National Governors Association Center for Best Practices & Council of Chief State School Officers, 2010; Silver, 1994; Yeap, 2009). Mathematical PP can be used before, during, and after problem solving (Silver, 1994; Yeap, 2009). Therefore, success in problem solving has been shown to be associated with success in PP and vice versa (Cai & Hwang, 2002; Silver & Cai, 1996). Furthermore, PP activities can promote students' conceptual understanding, foster their ability to reason and communicate mathematically, and capture their interest and curiosity (Cai et al., 2015; NCTM, 1991). It is thus natural to consider how PP can be integrated into mathematics curriculum and instruction.

In this chapter, we use a mathematical problem to illustrate what PP is and how to use the what-if-not strategy to pose problems. We then describe how to integrate PP into mathematics curriculum and instruction, in particular, we shall describe a lesson that applied the instruction model by Lv and Wang (2006). Finally, we shall report the results obtained from an experimental study with 56 seventh grade students in Macao, which indicated that the students' PP skills did improve when PP was integrated into the lessons.

2 What is Mathematical PP?

Problem posing involves the generation of new problems and questions aimed at exploring a given situation as well as the reformulation of a problem during the process of solving it (Silver, 1994). There are several types of PP tasks that have been identified in the literature on PP. Based on numerous works by Stoyanova (1998) and Silver (1994), Christou, Mousoulides, Pittalis, Pitta-Pantazi, and Sriraman (2005) described five types of tasks defined by the nature of the problem students are asked to pose: (i) a problem in general (free situations), (ii) a problem with a given answer, (iii) a problem that contains certain information, (iv) questions for a problem situation, and (v) a problem that fits a given calculation. Cai and

Jiang (2017) identified the following four types of PP tasks included in elementary mathematics textbooks used in China and the U.S.: (i) posing a problem that matches the given arithmetic operation(s); (ii) posing variations on a question with the same mathematical relationship or structure; (iii) posing additional questions based on the given information and sample question(s); and (iv) posing questions based on given information. They are ordered broadly from the most to the least mathematically constrained. Cai and Jiang (2017) found that (a) very low percentages of problems in both Chinese and U.S. textbooks are PP tasks; (b) the distributions of PP tasks were uneven across content areas and grade levels; and (c) the majority of PP tasks in the 2010s Chinese mathematics textbooks are of the third type, whereas the majority of PP tasks in the 2010s U.S. mathematics textbooks are of the first type. In the study by Hu, Cai, and Nie (2015), about 80% of the PP tasks are represented under the after-school practice and review sections in the Chinese 2000s mathematics textbook series. These studies on mathematics textbooks indicated that although the textbook developers in China and U.S. tried to incorporate PP into the mathematics curricula, the inconsistency in terms of distribution of PP tasks in different content areas, grade levels, PP types, and usage for learning and/or practice suggests a need for more effort to make it a classroom routine.

3 The Importance of PP in Mathematics Education

We shall look at the importance of PP from the following four aspects: (1) the mathematicians' experiences, (2) the objectives of mathematics education, (3) the implementation of mathematical PP in the Chinese mathematics education, and (4) the literature in mathematics education.

To formulate a mathematical problem is the first step in mathematics research. Einstein and Infeld (1938) stated that the formulation of the problem is often more essential than its solution. To them, the solution of a mathematical problem sometimes can be merely a matter of mathematical or experimental skills, whereas the formulation of the problem is a kind of creative work.

To develop students' mathematical PP skills has become one of the objectives of mathematics curriculum in different countries around the world. In the U.S., the NCTM (2000) argued that students must be given opportunities to "formulate interesting problems based on a wide variety of situations, both within and outside mathematics (p.258)". In the latest mathematics curriculum standards published by the Chinese Ministry of Education (2011), the objectives of mathematics education are stated from the following four aspects: knowledge and skills, mathematical thinking, problem solving, and attitudes. Out of the three points in the *knowledge and skills* aspect, two are related to mathematical PP. One is to formulate algebraic problems from real-world situations, the other is related to statistics and probability. To pose a problem is the first step of doing statistics. Out of the four points in the *problem solving* aspect, the first one is also related to PP. To learn mathematics is to learn how to pose problems from mathematical perspectives and to solve it using mathematical methods.

Problem posing has been included in the Chinese mathematics curricula for a long time (Cai & Jiang, 2017). When the first author studied in primary and secondary schools in 1978-1986, she had already been exposed to using PP tasks. Although she could not recall much of the things that happened in her primary and secondary education, what remains ingrained in her memories was her experience in PP. That experience was amazing and had helped her to better understand the mathematical structure embedded in the problems. In December 2016, when she visited a primary school in Hangzhou that houses a small mathematics library, she managed to find several mathematics textbooks published in the 1980s. Figure 1 shows two examples of exercises taken from two textbooks published in 1989 in China. In the first example, two pieces of information are given, students are then required to pose an addition and a subtraction problems. Under this situation, the students need to understand the meanings of the sum and the difference of two numbers before posing problems. In the second example, the givens are three pieces of information and the possible questions to ask for each. Strictly speaking, it may not be a PP task, however, it is a kind of linking item between problem solving and PP.

王华的妈妈在工厂里织🧦。上午，织大人袜 32 双，儿童袜 48 双。

（1） 口头提出一个要用加法算的问题，再算出来。

（2） 口头提出一个要用减法算的问题，再算出来。

Translation: Wang Hua's mother is a textile worker weaving socks. In a morning, she can weave 32 pairs of socks for adults and 48 pairs of socks for kids.

1) Orally raise a question that can be answered using addition. Then do the calculation.

2) Orally raise a question that can be answered using subtraction. Then do the calculation.

先把有关的条件和问题连起来，再计算。

3 个人做 15 面小旗，	还要做几面？
3 个人做小旗，每人做 5 面。	一共做多少面？
要做 15 面小旗，已经做了 5 面	平均每人做几面？

Translation: Matching the givens with the questions, and then answer them.

Three people have made 15 flags.	How many more flags need to be made?
Three people have made 5 flags each.	How many flags altogether?
15 flags need to be made. 5 have been made already.	On average, how many flags has one people made?

Figure 1. Problem posing tasks included in mathematics textbooks in China in 1980s

There is a kind of mathematical task called "*juyi fansan* (举一反三)" in China. This kind of task is usually used either after a topic has been taught or when a student fails to solve a problem in homework and examinations. "*Juyi fansan*" comes from a saying in the *Lunyu* (论语): 举一隅，不以三隅反，则不复也. The translation of this saying is roughly as follows: If a person is taught one aspect of an idea in a lesson and cannot point out three other aspects, then the lesson should not be repeated for the person. This translation does not do justice to the original saying. In traditional Chinese language, the character "三" means "many", therefore, "*juyi fansan*" can be taken to mean that by just demonstrating one example in teaching, students must then be able to apply the knowledge and skills learnt in that example to solve many similar problems. In other words, the example embodies a certain commonality found in similar problems. Only interpreting it in this way, *juyi fansan* is closely related to "*chulei pangtong*" (触类旁通). In Chinese language, *juyi fansan* and *chulei pangtong* are always taken as synonyms although *chulei pangtong* means extending a concept to another by analogical reasoning, which is generalization to a wider context. In doing *juyi fansan* tasks, students are asked to pose similar problems that they have just been taught in classes or when they make mistakes.

Problem posing potentially has a positive effect on concept learning, skill development, mathematical processes, metacognition and attitudes, all five aspects of the Singapore mathematics curriculum "pentagon" framework. For instance, PP has a positive effect on students' development of mathematical problem solving skills through helping them understand mathematical concepts better, develop mathematical skills, and problem solving processes (Cai et al., 2015). It may also help students to develop a more positive attitude towards mathematics and mathematics learning, as well as higher meta-cognitive skills (Chen, Van Dooren, & Verschaffel, 2015).

4 Mathematical PP Strategy: What-If-Not

Building on Polya's "looking back" stage in problem solving, Brown and Walter (1990) proposed the well-known *What-if-not* strategy for mathematical PP. Along the same lines, Abu-Elwan (2002) as well as Cai and Brook (2006) suggested posing problems through a process of extending or generalizing an already solved problem. Indeed, Gonzales (1998) even referred to this process as a fifth step to Polya's four-step model. Lavy and Bershadsky (2003) divided the use of *What-if-not* strategy into two stages. In the first stage, all the attributes included in the statement of the original problem are listed. In the second stage, each of the listed attributes is challenged by asking "what if not attribute k?" and alternatives are then proposed. Each of the offered alternatives creates a new problem situation. Figure 2 below demonstrates how Lavy and Bershadsky's (2003) model works using a Primary 1 mathematical problem.

The Add to 12 problem: Put 1, 2, 3, 4, 5, 6, and 7 into the circles (one number for one circle) so that the sum of three numbers on the same line will add up to 12.

Figure 2. The Add to 12 problem

The *Add to 12* problem is not difficult to solve. The central circle is shared by every strand, therefore, the six numbers to be filled in the remaining circles should satisfy the condition that every pair will add up to the same total. For this purpose, only 1, 4, and 7 could be put at the centre. However, only by placing 4 at the centre and pairing the largest of the remaining numbers with the smallest one, then the second largest one with the second smallest one (i.e., 1 and 7, 2 and 6, 3 and 5), can the totals on each line sum to 12. In this case, a very useful strategy is to match the largest one with the smallest one. We can list the attributes in the original problem as suggested by Lavy and Bershadsky (2003) below:

(1) There are three strands with one circle at the centre.
(2) There are three circles on each strand.
(3) Seven numbers (1, 2, 3, 4, 5, 6, and 7) are to be put into the circles with one number in one circle.
(4) The sum of the three numbers on the same line will add up to 12.

One of the attributes in the question above is the number of strands, which is three. We can ask "What if the number of strands is NOT three?" It can be four, five, or six! Then the numbers that need to be put into the circles can be changed to 9, 11, or 13 accordingly (see Figure 3). As long as the number at the centre is determined, the remaining numbers for the other circles can be worked out in similar ways as above.

Another possible attribute concerns the number of circles on each strand, which is three in the question above. What if the number of circles on each strand is NOT three now? It can be four (see Figure 4(a))! In this case, 10 numbers are required to be put into the circles.

We can also change the previous two attributes together. In Figure 4(b), the number of strands is changed to four, and the number of circles on each strand is four. In Figure 4(c), the number of strands is changed to five, and the number of circles on each strand is four.

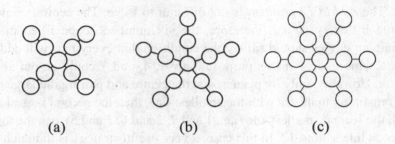

(a)　　　　　　　　(b)　　　　　　　　(c)

Figure. 3. To Change the number of strands to four, five and six.

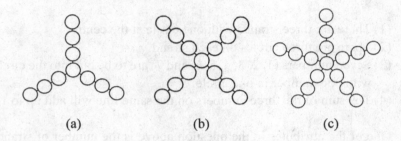

(a)　　　　　　　　(b)　　　　　　　　(c)

Figure 4. To change the number of circles and/or the number of strands.

Thirdly, if the numbers to be put into the circles are not 1 to 7, what kind of numbers can they be? The answer is any arithmetic number sequence. We can also modify this attribute as "If we add up all the numbers on the same strand, we can get the same results for all the strands without specifying what sum we can get". This may make the problem an open-ended one because it can have multiple solutions. For the newly reformulated problems, we can further discuss the solution methods with the students to see whether we can apply the similar matching strategy. The above generalization and problem solving processes help the students to see the deep structures embedded in the problem and to check whether a similar solution method can be applied or whether any modification is needed.

5 Integrating Mathematical PP into Instruction

Lv and Wang (2006) described a lesson structure that integrates mathematical PP into classroom. This lesson structure comprises the following four stages: (1) Create a mathematical situation; (2) Pose mathematical problems; (3) Solve mathematical problems; and (4) Apply the mathematical knowledge just learnt. What follows is a description of a lesson conducted by Ms. Wu Zhengxian, a primary school Mathematics teacher in Beijing who is now the director of Primary Mathematics Office at the Beijing Education Scientific Research Network. She has written about 15 books on primary mathematics education. In November 2012, she was invited to conduct an exemplary lesson on the topic of "Overlapping Problems" for Macau primary mathematics teachers. The topic is related to the computation of elements in two or more sets whose overlapping intersection is not an empty set.

Create a mathematical situation. Ms. Wu created the following situation for the class: A group of children are lined up. Counting from the front, Xiaoming was the fifth in the line. If counted from the end, Xiaoming was also the fifth in the line.

Pose mathematical problems. Ms. Wu asked the class to pose problems based on the above situation. Naturally all the students posed the problem: "How many children are lining up there?" Without hesitating, almost all the students gave the answer "10" which came from 5+5.

Solve mathematical problems. Ms. Wu led the class to solve the problem using different approaches including *Act it out* and *Draw a picture*. She followed the class' suggestion to draw 10 circles to represent the 10 children. She then counted from the front to the fifth circle and marked it, and counted from the end to the "Fifth", which was not the one that was just marked. Subsequently she asked the class how they would resolve the conflict. The class suggested erasing one circle, which she did, then followed by checking again to confirm that the two given conditions had been satisfied. Finally, she led the class to discuss how to write down a mathematical expression (i.e., 5+5-1) to represent the solution. Next, she asked the class to explain the meanings of the two 5s and 1 in the expression. As a result, they spent more than 10 minutes to discuss why the solution was not 5+5 but 5+5-1, and why they needed to take away 1.

Apply the mathematical knowledge just learnt. Ms. Wu gave students the following problem to solve so that students could apply what they had just learnt:

> In a class, there are 5 students participating in a Chinese competition, 9 students participating in a mathematics competition, and 2 participating in both. How many students are there participating in either Chinese or mathematics competition?

After giving students about three minutes to work on the problem, Ms. Wu discussed the multiple solutions, i.e., 5+9-2, (5-2)+9, 5+(9-2), and 3+7+2 using Venn diagrams, focusing on the counting principles (counting all without skipping and double counting). Here she used the "One Problem Multiple Solutions" approach which has been discussed by many researchers (e.g. Cai & Nie, 2007).

After implementing the four steps of the model proposed by Lv and Wang (2006), Ms. Wu further asked the class to make up stories for Figure 5.

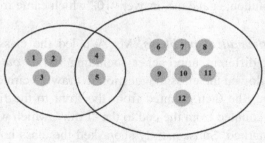

Figure 5. Make up a story for this figure.

Students made up stories like participating in competitions in different subjects, favourite TV programs, favourite fruits, and so on. At the end of the lesson, she posed the following open-ended problem, for which the answers can be any whole number between 1 and 5, to the class:

In a class, 5 students participated in a Chinese competition, and 7 students in the mathematics competition. How many students participated in both competitions?

Although this open-ended problem could not be fully discussed in the class, it was very encouraging to see quite a few students coming up to talk with Ms. Wu about their thinking and solutions.

This was a wonderful lesson integrating mathematical PP both in the beginning and at the end of the lesson. It started from a familiar experience students had. The problem looked so simple, but the answer of 10 was not correct. The whole class discussion did not only illustrate how to solve it, but also provided a clear explanation why it was necessary to take away 1. The teacher also applied the "One Problem Multiple Solutions" approach (Cai & Nie, 2007) in the fourth stage. The focus was shifted from multiple solutions to the more fundamental counting principle. The class ended with a mathematical PP activity and an open-ended problem, which extended the class to deeper thinking. We can conclude from their smiling faces that they had learnt what they were intended to learn. The most important thing is that they would continue their thinking about mathematics even after the class ended.

6 An Experimental Study

Aforementioned, PP can be integrated into problem solving by adding a fifth step after Polya's look-back stage (Abu-Elwan, 2002; Cai & Brook, 2006; Gonzales, 1998). The first author carried out a quasi-experimental study on the teaching of simultaneous linear equations in two variables to address the following two research questions:

(1) Compared with the traditional teaching method, can the teaching of problem solving through integrating PP improve students' abilities in mathematical problem solving?

(2) Compared with the traditional teaching method, can the teaching of problem solving through integrating PP improve students' abilities in mathematical PP?

6.1 *Method*

The experiment was conducted during four sessions of 45 minutes each.

Sample. The sample was two seventh grade classes with a total of 56 students from the same secondary school. One class with 16 boys and 12 girls, as the control group, adopted the traditional teaching method guided by Polya's four-stage problem solving theory. The other class with 18 boys and 10 girls, as the experimental group, used an instruction method incorporating PP after Polya's fourth stage. The study was conducted over four lessons on using simultaneous linear equations in two variables.

Instrument. Both pre-test and post-test were used to collect data from the two groups. The pre-test consisted of 14 items. Thirteen of them were problem solving items, and the other was a PP task on a linear equation in one variable. Item 1 was to solve simple simultaneous linear equations in two variables by the substitution method. Items 2-5 asked students to write expressions of a given statement like "y is 5 less than x". Items 6-9 were intended to examine students' translation and planning skills (Cai, 1995). Item 10 asked students to write a linear equation in one variable for a word problem. Items 11-13 asked students to formulate simultaneous linear equations in two variables. Item 14 asked students to write a word problem for $3x - 8 = 2x + 6$.

The post-test consisted of 8 items. Seven of them were problem-solving items on simultaneous linear equations in two variables, and the other was a PP task in simultaneous linear equations in two variables.

Data analysis. All the responses were marked by a Masters student in mathematics education. In the pre-test, the responses for the first ten were awarded either 1 point for correct answers or 0 point for incorrect answers. For items 11-13, 1 point was given for each correct equation built, therefore, the total possible points would be 2. For item 14, the responses were marked either 1 if the word problem can be solved using the given equation or 0 if it cannot. In the post-test, the responses for the first seven items were marked from two aspects, equation building and equation solving. For equation building, if a student can build one equation correctly, he will earn 1 point. If he can build two equations correctly, he will be given 2 points. Similarly, for equation solving, if a student can find the correct value for one variable, he will be given 1 point. If a student can find the correct values for two variables, he will be given 2 points. For item 8, the coding method used by Cai et al. (2013) was adapted. Interrater coders then judged whether the student made an attempt to pose a problem and whether the posed problem was situated in a realistic context. The posed problems were also coded according to whether they correctly represented none, one, or both of equations. Therefore, the possible scores for item 8 was 0 to 2.

Students' mathematical problem solving skills were measured by the total scores in the first 13 items in pre-test and the first 7 items in post-test. Their PP skills were measured by the total scores in the last item in both tests. There was only one PP item in the pre- or post-tests each, which is one of the limitations of the current study.

6.2 *Results*

The Cronbach Alphas for the pre-test and the post-test are 0.80 and 0.82 respectively. Both are within [0.8, 0.9], therefore, the test is statistically reliable (Nunnelly, 1978).

The means scores of both experimental and control groups are shown in Table 1. The t-test results indicated that there were not significant differences between the two groups in problem solving and PP in the pre-test. However, the experimental group performed significantly better than control group in PP in the post-test with a medium effect side ($d = 0.53$). The study indicated that the students could develop their mathematical PP skills as long as they are provided with the opportunities to learn. When looking at the equation building aspect in each item, we could find that the experimental group performed significantly better than the control group in three items, whereas the control group performed significantly better than the experimental group in two items. When looking at the equation solving aspect in each item, we could find that the experimental group performed significantly better than control group in two items, whereas the control group did not perform better than the experimental group in any of the seven items. It is important to bear in mind that the experiment lasted for only four sessions with a total of 180 minutes. Students should be able to learn more if the experiment was longer.

Table 1

Mean Scores (SD) of Experimental and Control Groups in both Pre- and Post-tests

	Experimental Group	Control Group	t	p
Pre-test				
Problem Solving	6.75 (4.00)	6.96 (4.06)	-.199	.843
Problem Posing	.04 (.19)	.14 (.36)	-1.406	.167
Post-test				
Problem Solving	18.50 (4.13)	17.21 (7.17)	.822	.208
Problem Posing	1.25 (.97)	.75 (.93)	1.974*	.027

$*p < .05$

7 Conclusion

This chapter focuses on how to integrate mathematical PP into mathematics instruction and problem solving. Our textbook analysis (Cai & Jiang, 2017) indicated that more effort is still needed to make mathematical PP a classroom routine. The experimental study reported above showed that integrating PP into the teaching of problem solving did improve students' PP skills even after four 45-minute sessions. If mathematical PP can be implemented in the mathematics classroom in the way demonstrated by Ms. Wu, learning mathematics will become a more enriching experience for many children.

Acknowledgement

This project was sponsored in part by the University of Macau (UM) Multi-Year Research Grant. We are grateful to UM for the support. Opinions expressed herein are those of the authors and do not necessarily represent the views of UM.

References

Abu-Elwan, R. E. (2002). Effectiveness of problem posing strategies on prospective teachers' problem solving performance. *Journal of Science and Mathematics Education in S. E. Asia, 1*, 56-69.

Brown, S. I., & Walter, M. I. (1990). *The art of problem posing* (2nd edition). Hillsdale, NJ: Lawrence Erlbaum Associates.

Brown, S. I., & Walter, M. I. (1993). Problem posing in mathematics education. In S. I. Brown, & M. I. Walter (Eds.), *Problem posing: Reflections and application* (pp. 16-27). Hillsdale, NJ: Lawrence Erlbaum Associates.

Cai, J. (1995). A cognitive analysis of U.S. and Chinese students' mathematical performance on tasks involving computation, simple problem solving, and complex problem solving. *Journal for Research in Mathematics Education Monograph Series, 7*. Reston, VA: National Council of Teachers of Mathematics.

Cai, J., & Brook, M. (2006). Looking back in problem solving. *Mathematics Teaching Incorporating Micromath, 196*, 43-45.

Cai, J., & Hwang, S. (2002). Generalized and generative thinking in U.S. and Chinese students' mathematical problem solving and problem posing. *Journal of Mathematical Behavior, 21*(4), 401-421.

Cai, J., Hwang, S., Jiang, C., & Silber, S. (2015). Problem posing research in mathematics: Some answered and unanswered questions. In F. M. Singer, N. Ellerton, & J. Cai (Eds.), *Mathematical problem posing: From research to effective practice* (pp. 3-34). New York, NY: Springer.

Cai, J., & Jiang, C. (2017). An analysis of problem-posing tasks in Chinese and US elementary mathematics textbooks. *International Journal of Science and Mathematics Education, 15*(8), 1521-1540.

Cai, J., Moyer, J. C., Wang, N., Hwang, S., Nie, B., & Garber, T. (2013). Mathematical problem posing as a measure of curricular effect on students' learning. *Educational Studies in Mathematics, 83*(1), 57-69.

Cai, J., & Nie, B. (2007). Problem solving in Chinese mathematics education: Research and practice. *ZDM Mathematics Education, 39*, 459-475.

Chen, L., Van Dooren, W., & Verschaffel, L. (2015). Enhancing the Development of Chinese Fifth-Graders' Problem-Posing and Problem-Solving Abilities, Beliefs, and Attitudes: A Design Experiment. In F. M. Singer, N. F. Ellerton, & J. Cai (Eds.), *Mathematical Problem Posing; From* Research to effective practice (pp. 309-329). New York: Springer.

Chinese Ministry of Education. (1986). *Jianguo yilai zhongxue shuxue jiaoxue dagang huibian (1949-1985)* [A collection of mathematical syllabuses (1949-1985)]. Beijing, China: Author.

Chinese Ministry of Education. (1992). *Yiwu jiaoyu quanrizhi xiaoxue chuji zhongxue kecheng jihua* [Curriculum plan for full-time primary and middle schools of the compulsory education]. Beijing: Author.

Chinese Ministry of Education. (2000a). *Jiunianzhi yiwu jiaoyu quanrizhi xiaoxue shuxue jiaoxue dagang (shiyong xiuding ban)* [Mathematics syllabus for full-time primary schools of nine-year compulsory education (Revision of the trial version)]. Beijing: Author.

Chinese Ministry of Education. (2000b). *Jiunianzhi yiwu jiaoyu quanrizhi chuji zhongxue shuxue jiaoxue dagang (shiyong xiuding ban)* [Mathematics syllabus for full-time middle schools of nine-year compulsory education (Revision of the trial version)]. Beijing: Author.

Chinese Ministry of Education. (2001). *Quanrizhi yiwu jiaoyu shuxue kecheng biaozhun (shiyan gao).* [Curriculum standards for school mathematics of nine-year compulsory education (trial version)]. Beijing, China: Beijing Normal University Press.

Chinese Ministry of Education. (2011). *Quanrishi yiwu jiaoyu shuxue kecheng biaozhun.* [Mathematics curriculum standard of compulsory education (2011 version)]. Beijing: Beijing Normal University Press.

Christou, C., Mousoulides, N., Pittalis, M., Pitta-Pantazi, D., & Sriraman, B. (2005). An empirical taxonomy of problem posing processes. *ZDM Mathematics Education, 37*(3), 149-158.

Einstein, A., & Infeld, L. (1938). *The evolution of physics: the growth of ideas from early concepts to relativity and quanta.* New York: Simon and Schuster.

Gonzales, N. A. (1998). A blueprint for problem posing. *School Science and Mathematics, 98*, 448-453.

Hashimoto, Y. (1987). Classroom practice of problem solving in Japanese elementary schools. In J. P. Becker, & T. Miwa (Eds.), *Proceedings of the U.S.-Japan Seminar on Mathematical Problem Solving* (pp. 94-119). Carbondale, IL: Southern Illinois University.

Healy, C. C. (1993). *Creating miracles: A story of student discovery.* Berkeley, CA: Key Curriculum Press.

Hu, D., Cai, J., & Nie, B. (2015). Mathematics question posing and curriculum evolution: Comparison between two editions of primary mathematics textbooks. *Curriculum, Teaching Material, and Method, 35*(7), 75-79.

Keil, G. E. (1967). Writing and solving original problems as a means of improving verbal arithmetic problem solving ability. *Dissertation Abstracts International, 25*(12), 7109.

Kilpatrick, J. (1987). Problem formulating: Where do good problems come from? In A. H. Schoenfeld (Ed.), *Cognitive science and mathematics education* (pp. 123-147). Hillsdale, NJ: Erlbaum.

Lavy, I., & Bershadsky, I. (2003). Problem posing via "what if not?" strategy in solid geometry-a case study. *Journal of Mathematical Behavior, 22*, 369-387.

Lv, C., & Wang, B. (2006). Zhongxiaoxue shuxue qingjing yu tichu wenti jiaoxue yanjiu [*Research on mathematics teaching through using mathematical situations and posing problem in high school and primary school*]. Guiyang, China: Guizhou People's Publishing House.

National Council of Teachers of Mathematics (NCTM). (1991*). Professional standards for school mathematics*. Reston, VA: Author.

National Council of Teachers of Mathematics (NCTM). (2000). *Principles and standards for school mathematics*. Reston, VA: Author.

National Governors Association Center for Best Practices & Council of Chief State School Officers. (2010). *Common Core State Standards for Mathematics*. Retrieved from http://www.corestandards.org/math.

Nunnelly, J. C. (1978). *Psychometric theory* (2nd ed.). New York: McGraw-Hill.

Silver, E. A. (1994). On mathematical problem posing. *For the Learning of Mathematics, 14*(1), 19-28.

Silver, E. A., & Cai, J. (1996). An analysis of arithmetic problem posing by middle school students. *Journal for Research in Mathematics Education, 27*, 521-539.

Singer, F. M., Ellerton, N., & Cai, J. (2013). Problem-posing research in mathematics education: New questions and directions. *Educational Studies in Mathematics, 83*(1), 1-7.

Singer, F. M., Ellerton, N., & Cai, J. (2015). *Mathematical problem posing: From research to effective practice*. Springer.

Stoyanova, E. (1998). Problem posing in mathematics classrooms. In A. McIntosh, & N. F. Ellerton (Eds.), *Research in mathematics education: A contemporary perspective* (pp. 164-185). Perth, Australia: Edith Cowan University.

van den Brink, J. (1987). Children as arithmetic book authors. *For the Learning of Mathematics, 7*, 44-48.

Wang, B., & Lv, C. (2000). Chuangxin yu zhongxiaoxue shuxue jiaoyu [Innovation and mathematical education of the primary and middle school]. *Journal of Mathematics Education, 9*(4), 34-37.

Yeap, B.H. (2009). Mathematical problem posing in Singapore primary schools. In B. Kaur, B.H. Yeap, & M. Kapur (Eds), *Mathematical problem solving (*pp. 102-116*)*. Singapore: World Scientific.

Chapter 13

A Vicennial Walk Through 'A' Level Mathematics in Singapore: Reflecting on the Curriculum Leadership Role of the JC Mathematics Teacher

Weng Kin HO   Christina RATNAM-LIM

Teachers are curriculum leaders. Many 'A' Level Mathematics teachers, however, perform day-to-day enactment of the mathematics curriculum without even realizing that they are leaders in this respect. With an audience of 'A' Level teachers in mind, this chapter aims to create and promote an awareness of their curriculum leadership role. A framework of levels of curriculum decision-making is used to show that, contrary to practitioner perceptions, teachers and heads of departments have an active role to play in curriculum leadership. A major part of this chapter shows the relationship between Contexts 1 and 2 of the framework, mapping out how policy decisions made at the national level affected programmatic (discipline) decisions, resulting in major changes in the 'A' Level Mathematics Syllabus over the past two decades. We hope that the 'A' Level Mathematics teachers reading this chapter are then able to appreciate the purpose for the changes in the mathematics syllabus and their implications on curriculum practices in the classroom. Having attained a better understanding of their important roles as active curriculum leaders, rather than passive deliverers of the mathematics syllabus, 'A' Level Mathematics teachers can then actively formulate goals, design tasks and implement classroom activities that together constitute the experienced 'A' Level Mathematics Curriculum.

1 Introduction

Teaching, as a profession, demands that teachers continually exercise their professional judgement to explore, develop, and select innovative designs that engage their students in learning. In that sense, teachers are curriculum leaders who exercise active agency in planning, designing, monitoring and reviewing their lessons, material, and assessments (Hairon, Tan, Lin, & Lee, 2017). In reality, however, the constraints of a hectic teaching schedule and demands of a high-stakes examination schooling system curtail such professional agency. Critical curriculum decisions—such as what knowledge is of most value and worth, what activities and experiences could help students attain this knowledge, and how these experiences are organized and sequenced for effective learning (Eisner, 2002; Schwab, 1973; Tyler, 1949)—are directed by the examination syllabus, not by the practitioner in the classroom. It is not surprising, then, that teachers do not see their role as *curriculum leaders* in their daily practice. Instead, curriculum decision-making tends to be centralized at the level of the Ministry of Education (MOE), which publishes the syllabus for each subject.

To exacerbate matters, "nationwide educational policies and movements are made frequently and within a short space of time from each other" (Ratnam-Lim, 2017, p. 45). To many 'A' Level Mathematics teachers, experiencing three to four major changes in the 'A' Level Mathematics syllabus within a short span of fifteen years is seen as 'turbulent' and frustrating, especially when they do not see the purpose and need for the changes. New inclusions of topics into the syllabus will imply that something else must be displaced from it (e.g., the exclusion of approximation of binomial distribution by normal distribution *makes space* for the inclusion of the topic of discrete random variables, probability distributions, expectation and variance). Indeed any changes or teaching innovations which are not matched by a corresponding change in teachers' belief can hardly be sustained (Guskey, 2002; Thomas, 2014). Conversations with many JC mathematics teachers (here and hereafter, we use the term 'JC mathematics teachers' to include generically all mathematics teachers who are preparing students to take 'A' Level

Mathematics) revealed that while they could understand that changes in the curriculum were meant for a 'greater' good, they felt that there was little they could do once the change in the syllabus has already been announced (Ho, Toh, Teo, Zhao, & Hang, 2018).

The point we make here is that teachers need not view themselves as passive civil servants responsible for carrying out the implementation as spelt out in the *intended* curriculum designed by MOE. The difference teachers make to education is the way the intended curriculum is *enacted* through their active participation in designing, playing out and reviewing their understanding of the intended curriculum (Eisner, 2002; Schwab, 1973; Thompson & Pascal, 2012). Ultimately, this enactment brings about the learned curriculum, i.e., the body of knowledge and skills which are eventually acquired by the students. For a shift of the teacher's belief from being a passive implementer of the intended curriculum to being an active enactor of the curriculum, there must first be an appreciation of 'bigger' ambient changes at the various levels of curriculum decision making (Malthouse, Roffey-Barentsen, & Watts, 2014; Thompson & Pascal, 2012).

In this chapter, we take on the challenge of unpacking the trends in the curriculum changes that have occurred in the 'A' Level Mathematics over the last two decades, making sense of the motivation behind them and underscoring their implications. We aim to paint a panorama of the 'A' Level Mathematics education landscape so as to create and promote awareness among JC mathematics teachers that they are not merely passive deliverers of the intended curriculum 'handed down' by MOE, but in fact, have a part to play as *leaders actively shaping and enacting the curriculum to produce the experienced curriculum.*

To do this, we have used a framework of levels of curriculum decision making to show the various contexts in which curriculum decisions are made. The aim of this section is to help JC mathematics teachers see that, contrary to practitioner perceptions, teachers and heads of departments have an active role to play in curriculum decision making. This chapter also explicates the relationship among the four quadrants of the framework, mapping out how policy decisions made at the national level affected programmatic (discipline) decisions, resulting in major changes in the 'A' Level Mathematics Syllabus over the past two decades. More importantly, we hope to foreground how the practitioner in the field exercises professional leadership and agency in realizing the intentions espoused at the policy and programmatic levels.

2 Contexts of Curriculum Decision-Making

Curricula are the result of decisions made by educators based on their best understanding of the needs of learners and the socio-political milieu (Eisner, 2002). The decision-making process is difficult, and fraught with dilemmas and tensions. Considerations for what to include (goal, topic, activity, task, question, component, material, etc.) would also involve what to exclude. JC mathematics teachers are often torn between covering the intended curriculum spelt out in the syllabus and engaging students in deeper understanding of the concepts (e.g., the choice between teaching the conditions for applying the Central Limit Theorem to approximate the distribution of sample mean and teaching the reasons that justifies this approximation). Dilemmas of these kinds are endless. Curriculum decisions are therefore not neutral, but are influenced by the context in which they are made.

Eisner (2002) provided us two sets of dilemmas that occur in making curriculum decisions: (1) scale and scope of our attention (from the particular to the general), and (2) the time at which and for which curriculum decisions are made (for the present or in the long term). The space spanned over the two axes (see Figure 1) yield four possible contexts of curriculum decisions. While the resulting framework yields a useful way for discussing curriculum decisions, caution should be taken in regarding the resulting quadrants as neatly distinct as the diagram may imply. Descriptions of examples in each quadrant are dependent on the socio-cultural context. The Singaporean 'A' Level Mathematics examples we describe in each quadrant would be different from the examples Eisner described based on the American context. For better reading, we organize the aforementioned contexts and their corresponding examples as follows. In Section 3, we elaborate on Context 1, and in Section 4, Context 2. Then we group the examples of Contexts 3 and 4 together in Section 5.

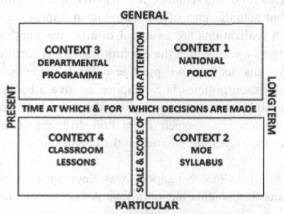

Figure 1. Contexts of Curriculum Decisions

3 Context 1

Curriculum decisions in Context 1 affect the masses (general) and are intended for the long term. Decisions in this context typically involve national policy concerns for education. Eisner (2002) noted that "such policies do not exhaust the decisions that need to be made, nor are they adequate for operating programs within schools, but they do establish the directions and boundaries for other decisions" (p. 28). That is, the actions and considerations in Context 1 can have cascading effects into the other contexts—the policy decisions and broad cultural influences that affect the nation will necessarily affect specific subjects, programmes and classrooms. The directions and boundaries emanating from Context 1 influence the pathways and trajectories of the nation's education system. In the context of the Singaporean education scene, it is perceived that nationwide policies and movements are made frequently and in a short space of time from each other. For example, the *Teach Less, Learn More* (TLLM) movement came barely eight years after the *Thinking Schools, Learning Nation* (TSLN) movement. It is crucial, therefore, for curriculum leaders to closely examine the policies, movements and slogans emanating from Context 1 and interpret the implications for the department and classroom. Upon close examination, the curriculum leader may find that these initiatives and movements are usually not singular events in themselves, but actually part of a continuum in improving teaching and learning. Such realizations are essential to allay teachers' anxieties and general feeling of wariness that the education sector is constantly buffeted by change. In this section, we provide a brief historical overview of changes in education policies in Singapore, to give a background to the subtle shifts in curriculum orientations over the years. In so doing, we hope teachers get a sense of how such orientation shifts resulting from policy changes affect their teaching practice in the classroom.

From 1819 to 1963, Singapore was developed as part of Great Britain's Straits Settlements in South East Asia. One of the legacies left by the British to Singapore's education system is the use of the Higher School Certificate in Education (HSCE) Examinations set by the University of Cambridge Local Examination Syndicate, UCLES (established since 1858), by the University of Cambridge as certification of completing "Higher School", and as a matriculation examination for entrance into universities in Singapore or Britain. This examination is now known as the 'A' (Advanced) Level Examination undertaken by most students in junior colleges in Singapore. It is not surprising that the curriculum orientation for the education system in this context of preparation for study in the University, would be *scholarly academic*. Scholarly academic ideologists view curriculum creation from the perspective of the academic disciplines (Schiro, 2013). The curriculum content that is valued is derived from the body of shared knowledge collected within the academic disciplines found within institutes of higher learning (universities). For instance, all students at pre-university level should learn about the solution of simple trigonometric equations, small angle approximations, cumulative distribution function, and theoretical and empirical interpretations of probability, etc. These are the "the distinctive disciplines of knowledge" (Whitefield, 1971, p. 12) that are deemed worthwhile for learning.

The period between 1959 and 1978, in which Singapore struggled to find its feet as a fledgling independent nation, has often been termed the "Survival Phase". In the process of ensuring the political and economic survival of the small city-state, important policies in the field of education focused on "an intimate link between education and economic development", resulting in "developing new skills and work attitudes to accommodate new economic strategies" (Goh & Gopinathan, 2008, p.14). The subjects that were valued (even till today) were English language (the language of commerce), mathematics, science, and technical subjects. Till today, all educational policies and reforms are undertaken with the nation's economic growth and sustainability as priority. This ushered a subtle shift from the scholarly academic orientation to social efficiency. Curriculum workers holding the social efficiency orientation believe that the purpose of schooling is to meet the societal demands by training its youth to function as future mature contributing members of society (Schiro, 2013). Workplace skills and procedures developed by industry guarantee productive lives and ensure the perpetuation of a functional society (Bobbitt, 2009). For example, workplace mathematics would include financial mathematics, engineering mathematics, and so on. The social efficiency orientation became more obvious in the next phase in the education landscape.

As seen so far, MOE tends to engage in regular reviews of the education sector. There has been recognition since the 1990s by MOE that to survive and prosper in the future—particularly within the fast-changing globalised world of the 21st century—the education system would need to develop dispositions of innovation and flexibility (Lim-Ratnam, Atencio, & Lee, 2016). MOE has embarked upon a raft of curricular reforms aimed at enhancing student learning, particularly along the lines of lifelong learning, critical thinking, innovation, and positive attitudes and values (Deng, Gopinathan, & Lee, 2013; Ng, 2005; Teo, 2000).

Responding to the clarion call of *Thinking Schools, Learning Nation* (MOE, 1997), the Junior college Education Review and Implementation (JERI) committee was set up to bring about a revamp of the pre-university education curriculum in junior colleges and centralized institutes. Two major shifts were evident in this period: the use of Information Technology (IT) in the Singapore classrooms through different phases of IT Masterplans, and emphasis on quality and choice in learning.

MOE initiated the use of IT in the Singapore classrooms through different phases of IT Masterplans. The first phase (IT Masterplan I, i.e., MP1) 1997–2002 laid a strong ICT (Information Communication Technology) foundation in schools. This phase started with the setting up of basic infrastructures for computer-related facilities and IT-related training of teachers. The most impactful manifestation of MP1 in 'A' Level Mathematics curriculum was the introduction of Graphing Calculators (GCs)—hand-held scientific calculators that allow the user to plot graphs in Cartesian form and Parametric form, compute terms of sequences and finite series, solve polynomial and non-polynomial equations, and perform other tasks with variables or simple data structures like lists. We will elaborate more on the impact of this on the syllabus in the next section.

TSLN also brought about a shift of emphasis from efficiency-driven education system to one that focuses on quality and choice in learning. Along the same vein, *Teach Less, Learn More* (TLLM) from 2005 advocated that educators could better engage students in their own learning through more effective pedagogies. As such, teachers need to spend more time reflecting about their own classroom practices, constantly improving on the style and quality of interactions. TLLM reinforced the reduction of content knowledge to create space for more effective and engaging learning. JERI recommendations directly impacted the new 2006 'A' Level syllabi. For example, JC students are now expected to read a subject outside of their main area of specialization. This notion of offering contrasting subjects was instrumental in broadening the students' learning experience. Such initiatives signal a subtle shift from the social efficiency orientation to a learner-centred orientation, which allows for a more holistic development of the individual's intellectual, social, emotional and physical attributes (Schiro, 2013).

The shift to the learner-centred orientation was made more obvious in the most recent goal for education first initiated by Prime Minister Lee Hsien Loong in 2010 and 2011. To provide each child with the chance to develop the whole person. The slogan *"every school is a good school"* which was formalized by Minister of Education, Mr. Heng Swee Keat at the Work Plan Seminar 2012 was a deliberate step to counter elitism—a deep-seated indulgence in the belief that only good schools can develop good students. But what really makes a good school? MOE describes this through four 'every's:

(i) *Every school a good school.* A good school cares for its students, studying and knowing the needs, interests and strengths of her students and motivates them to learn and grow. This call for every school to be a good school subsequently unpacked into three further aspects (ii) – (iv) which we list below, and each of these three areas was the focus of Work Plan Seminars from 2013 to 2015.

(ii) *Every student an engaged learner.* A good school ensures all students acquire strong fundamentals of literacy and numeracy and develops them holistically, in character, knowledge and critical competencies. A good school creates a positive school experience for each student, making him a confident and lifelong learner.

(iii) *Every teacher a caring educator.* A good school has caring and competent teachers who are steadfast in their mission to impact lives.

(iv) *Every parent a supportive partner.* A good school has the support of parents and the community, working together to bring out the best in our children.

Table 1 gives an overview of the historical contexts and how they reflected particular curriculum orientations. More importantly, it also gives examples of the implications of these shifts in orientation on classroom instruction of Mathematics in Singaporean junior colleges. In Section 4, we shall map out the various policy changes that have taken place in Context 1 in Singapore since the later part of the 1990s, and attempt to show how decisions made in Context 2 are attempts at alignment with Context 1.

Table 1
Context and curriculum orientation with some implications for Mathematics JC Classroom

Period	Context & Curriculum Orientation	Implications for the JC Mathematics Classroom
1819-1963	British Straits Settlement; Matriculation for entrance into Universities (GCE 'A' Levels); scholarly academic orientation.	More emphasis on acquisition of content knowledge. E.g., (1) proofs of standard results (such as De Moivre's Theorem) need to be learnt by heart, (2) abstract concepts such as the algebra of sets, program flow chart, etc.
1959-1978 1979-1996	Survival phase & Quality education phase: economic development, workplace skills; social efficiency orientation.	More emphasis on applied mathematics, e.g., 3D trigonometry for modelling real life objects, use of differential equation to population dynamics, etc.
1997 - present	IT Masterplan I-II & Smart Nation.	Use of Graphing Calculators in teaching and learning mathematics, e.g., graphs of functions (Cartesian and parametric forms) are obtained using GC, instead of computation by hand.
	TSLN, TLLM, 21st Century Competencies, Every school a good school: learner-centred approaches (engaging the learner).	Learning experiences provide immersive environment for exploring topics in a more in-depth manner.

4 Context 2

The curriculum decisions emanating from the second quadrant focus on achieving long-term goals within a 'particular' context such as the syllabus for a particular subject. Each syllabus focuses on a particular subject, providing the long-term goals and outcomes for the learning of the subject. At the JC level, there are even different curricula offered within each subject, targeting three different levels: H1, H2 or H3 level. The syllabus for each 'H' Level details specific aims, learning objectives and assessment objectives. H2 subjects are somewhat comparable to the previous 'A' Level subjects in terms of intellectual demand. H1 subjects could be perceived as half of their H2 counterparts in terms of curriculum time. Specific H3 programmes are tailor-made to allow academically exceptional students to pursue that subject or area in which they have passion and aptitude. Unlike the previous Special Papers (commonly known as 'S' Papers), H3 programmes run on a separate syllabus which go beyond the H2 syllabus. The majority of the 'A' Level students sit for examinations under the H2 Mathematics Syllabus 9740.

In this section, we show how major changes of the 'A' Level Mathematics syllabi (Context 2) happened *in relation to* some of the policy changes (Context 1). Table 2 maps out how each policy decision influenced the shaping of the A-level Mathematics syllabus since the 1990s.

Content reduction. As can be seen in section 2.1, the policy changes emanating from TSLN (1997), IT Masterplans (1997-present) and TLLM (2005) would require changes to the syllabi (Context 2) to make space for the use of IT and greater emphasis on critical and creative thinking and learning skills required for the future. A major step in that direction must involve the reduction of content knowledge to be learnt. This reduction would free up more time for teachers and students to engage in activities that develop the aforementioned skills.

This content reduction in the 'A' Level Mathematics curriculum manifested in the Mathematics 9233 syllabus, which took its final form in 2001 after its 1999 interim version phased out. Further Mathematics 9234 was removed in 2005, aligning with the direction towards a more broad-based education. Expert knowledge from the syllabi review and steering committees advised which topics in Further Mathematics needed to be shifted to the H2 Mathematics Syllabus 9740, and which existing topics in Mathematics Syllabus 9233 were to be removed (and in some instances, moved to 'O' Level Additional Mathematics) to create sufficient space for

Table 2

Major initiatives since 1990s which resulted in changes in 'A' Level Mathematics syllabus

Initiative	General description	Syllabus change
TSLN (Thinking Schools, Learning Nation), 1997	Life-long learning, collective tolerance for change, schools as learning organisations, students develop both lower and higher thinking skills and processes.	Mathematics 9205 changed to Mathematics 9233 (Interim), 1999.
IT Masterplan 1, 1997	Equip schools with IT hardware, LCD projector in every classroom, whole school networking, IT use in 30% curriculum time.	GCs were introduced in 'A' Level Mathematics and used in the 2001. Further Mathematics 9234 (Revised) examination, which maintained to be "GC-neutral". Mathematics 9233 (Interim) changed to 9233 (Revised), 2001.
Review of JC/Upper Sec. Education, 2002	Reviewing at, both the macro and the micro levels, the curricula of all subjects for JC.	Parallel to this is the development of the new 2006 'A' Level curriculum, which was completed by 2004.
IT Masterplan 2, 2002	Integrate IT with curriculum design, student-centred learning environment, evaluation of the use of IT in education.	GCs are used for teaching and learning as well as assessment, i.e., in the H2 Mathematics (9740) examinations.
TLLM (Teach Less, Learn More), 2005		New 'A' Level curriculum, 2006 (independent of TLLM) - Broader and more flexible JC curriculum; - Subjects offered at H1, H2 and H3 levels; - Contrasting subjects at H1 and H2 levels; - Mathematics 9233 (Revised) changed to H2 Mathematics 9740; - Removal of the subject Further Mathematics 9234.
IT Masterplan 3, 2009	Transforming the learning environment, a continuum of MP1 and MP2.	
SMART Nation, 2014	Using new technologies to develop sustainable innovations, to run the place better, making a difference to people's lives.	A re-emphasis on STEM education
ESGS (Every School a Good School), 2012-15	Four every's in MOE education initiatives: - Every School a Good School; - Every Student an Engaged Learner; - Every Teacher a Caring Educator; - Every Parent a Supportive Partner.	- H2 Mathematics 9740 to 9758; - 21st century competencies (e.g., Critical and inventive thinking); - Re-emphasis on STEM education; - Focus on "disciplinarity" understanding of mathematics; use of constructivist pedagogies; - Re-introduction of the subject Further Mathematics 9649.

the additional topics. Strictly speaking, the change from 9233 to 9740 involved more intricate re-structuring of topics and most importantly, coherence of the new syllabus was ensured to the best of efforts. For example, a substantial removal of trigonometry from the revised 2001 'A' Level Mathematics Syllabus (9233) (e.g., general solution of trigonometrical equations, small-angle approximations) is a significant evidence that MOE was beginning to value the higher-order cognitive processes.

Content adjustments were also made to accommodate the use of IT in the classrooms. The aim was to realize ICT-enabled lessons, targeted up to 30% of curriculum time. In response to this, the new Mathematics 9233 (Revised) Syllabus included, for the first time, the aim for students to "integrate information technology to enhance the mathematical experience" (MOE, 2005). Emphasis was placed on solving real-world problems, and activities including communication about mathematical problem solving, and interpreting of the solution within context. As mentioned in Section 3, the most impactful manifestation of MP1 in 'A' Level Mathematics curriculum was the introduction of Graphing Calculators (GCs). GCs replaced traditional statistical tables in that they can compute probabilities related to Binomial, Poisson and Normal Distributions. GCs can also calculate p-values in hypothesis testing. In 2001, GCs were allowed in state examinations for the Further Mathematics 9234 (Revised) Syllabus, where it was emphasised that question set were "GC-neutral", i.e., students who did not use GCs would not be disadvantaged.

The second phase (IT Masterplan 2, i.e., MP2) 2003–2008 demanded a deeper integration of IT into lessons with the aim of increasing interactivity and engagement. New alternative pedagogies favoured by MP2 include the use of blogs, wikis, podcasts, e-portfolios, animations, video productions as well as mobile learning. In this second phase of IT Masterplan, GCs played an indispensable role in teaching and learning. Indeed, GCs are integral with regards to assessment since the H2 Mathematics 9740 Syllabus dictates that "the examination papers will be set with the assumption that candidates will have access to GC. Generally, unsupported answers obtained from GC are allowed unless the question states otherwise. … For questions where graphs are used to find a solution, candidates should sketch these graphs as part of their answers. … if there is written evidence of using GC correctly, method marks may be awarded" (MOE, 2006, p. 3).

In summary, the 'A' Level Mathematics syllabi experienced two rounds of content adjustment as it moved from Mathematics 9205 through 9233 (Interim) and 9233 to 9740 within a short span of 9 years. All topics that were removed, relocated, added or rewritten for clarity in at each syllabus revision are listed in Table 3.

Table 3

Content adjustments in 'A" Level Mathematics syllabi (1997 – 2017)

Status	9205 to 9233 (Revised)	9233 (Revised) to 9740	9740 to 9758
Remove	- Cumulative distribution function, F - Theoretical and empirical interpretations of probability	- Use of relation $(fg)^{-1} = g^{-1}f^{-1}$. - Topic of partial fractions has been moved to Additional Math - Explicit rules involved in inequalities, e.g., switching the direction of inequality when multiplying/dividing by negative numbers; modulus functions and inequalities - Logarithms and exponential functions; moved to Additional Mathematics - Use of binomial theorem to expand $(a + b)^n$ for positive integer n has been moved to Additional Math - Proofs by induction concerning topics outside sequence and series - 3D trigonometry removed; use of sine and cosine formulae moved to Additional Math - Trigonometrical functions moved to Additional Math - Application of de Moivre's theorem to prove trigonometric identities - The idea of a limit and the derivative as a limit. - The gradient of a tangent as the limit of the gradient of a chord - Problems involving small increments and approximation - Approximation of area under a curve using the trapezium rule	- relationship between three planes - geometrical effect of conjugating a complex number and of adding, subtracting, multiplying, dividing two complex numbers - loci of complex numbers - use of de Moivre's Theorem to find the powers and nth roots of a complex number - finding the volume of revolution about x- or y-axis when the curve is defined parametrically - normal approximation to binomial - normal approximation to Poisson - random, stratified, systematic and quota samples - tests for a population mean based on a sample from a normal population of an unknown variance for small sample sizes - use of t-test

(continued)

(continued)		- No mention of Karnaugh maps as tools to determine elementary probabilities - Concept of discrete and continuous random variables; probability density and distribution functions - Calculations of mean and variance for probability distributions other than Binomial, Poisson and Normal - Find by integration a general form of solution for a first order differential equation in which the variables are separable Find the general solution of a first order linear equation by means of an integrating factor - Numerical methods - Mechanics option - Formulate hypothesis testing concerning a population proportion, and apply a hypothesis testing using a normal approximation to a binomial distribution	
Relocate	- Basic probability laws of inclusion-exclusion; conditional probability; independent events; shifted from Pure Mathematics Option (c) to Probability	- Curve sketching is subsumed under Graphing Techniques - Arithmetic and geometric series is raised from the level of subtopic to topic - Proof by induction only involves sequences and series - Permutation and combination is grouped under Statistics	- Arithmetic and geometric progressions placed under sequences and series, not as a separate theme/topic - Permutation and combination subsumed under the topic of probability
Add	- Syllabus 9233 is arranged according to identified themes or topics	- Use of GC to graph a given function - Included the hyperbola - Finding the numerical solution of equations using a GC - Finding the numerical value of a derivative at a given point using a GC - Finding the numerical value of a definite integral using a GC - Use of t-test - Correlation coefficient and linear regression	- Integration techniques include integrands of product of sines and cosines - Concept of discrete random variables, probability distributions, expectation and variance
Rewritten for clarity		- Maclaurin's series: concepts of 'convergence' and 'approximation'	- Explicit mention of the concept of the absolute value function, and its properties with regards to lower and upper bounds - Sequence as a function of positive integer

Shifts in orientation. Subtle but gradual shifts in orientation have also been made in the 'A' Level Mathematics curriculum from scholarly academic, through social efficiency to learner-centred. The reduction in content and insertion of IT described earlier showed how Mathematics at 'A' Level had more obviously steered towards the social efficiency orientation, where applicability of mathematics in the real-world, e.g., at the workplace, is emphasised.

The shift to more learner-centred approaches is most evident in the 2016 'A' Level Mathematics Syllabus (H2 Mathematics 9758). An expanded suite of syllabi, with H2 Further Mathematics, was introduced to give students more options to choose from, thus catering better to their diversified needs. A new focus was placed on mathematical processes such as mathematical reasoning, mathematical modelling and communication. Learning experiences stated in the syllabus are instituted to positively influence the ways teachers teach and how students learn so that curriculum objectives can be achieved. It is stated explicitly in the revised syllabus that teachers are also encouraged to use pedagogies that are constructive in nature. The shift of the curriculum orientation from scholarly academic towards learner-entered is articulated in the aims of the new syllabus, where we see an increasing focus on the experience of the learner:

[9205] develops further the mathematical knowledge of students in a way that encourages *confidence* and provides understanding and *enjoyment*;

[items (1)-(2), 9233] develop further their understanding of mathematics and mathematical processes in a way that encourages confidence and enjoyment; develop a positive attitude to learning and applying mathematics;

[9740] develop positive attitudes towards mathematics.

Preparing the students towards a STEM-based education and catering for a wider variety of students' needs and interests, MOE expanded the 'A' Level syllabi to include Further Mathematics. The H2 Further Mathematics Syllabus 9649 is tailor-made for "students who are mathematically-inclined and who intend to specialise in mathematics, sciences or engineering or disciplines with higher demand on mathematical skills. It extends and expands on the range of mathematics and statistics topics in H2 Mathematics and provides these students with a head start in learning a wider range of mathematical methods and tools that are useful for solving more complex problems in mathematics and statistics" (MOE, 2017, p. 2).

The reader is alerted to two new components featured in the 2017 H2 Mathematics and Further Mathematics syllabi: one is set in the 'A' Level examinations and the other is to augment the summative assessment. *Problem in Real World Contexts* (PRWC) is set in examinations, where real-world situations are mathematised via mathematical models. Questions are then set to scaffold students to solve the given real-world problem within the context. Note that PRWC had already started in 'O' Level Mathematics recently. The other augmenting component is called *Learning Experience*. Here, students engage with meaningful discourse in which they actively reason about their understanding and communicate their ideas to their teachers and peers.

5 Contexts 3 and 4

5.1 *Context 3*

The third quadrant characterizes those curriculum decisions that have a general scope but with immediate or short-term goals. We have seen that the 'A' Level Mathematics syllabus as drawn out by MOE provides the long-term aims and outcomes of the intended curriculum for the particular 'level' (whether H1, H2 or H3). JC mathematics departments have the job of interpreting the intended curriculum and translating it into the enacted curriculum within the context of the two-year JC time frame. Such a translation typically involves coming up with a scheme of work that details the teaching and learning roadmap of the students, the scope and sequence of coverage of topics, and the assessment instruments spread along school terms.

It is in Context 3 where the heads of the mathematics department play a key role in supporting curriculum development. For JC mathematics departments, it is very common for teachers to meet regularly during 'contact time' to talk about each teacher's progress along the designated scheme of work, exchange pointers concerning students' learning difficulties, and discuss test and examination question-items as well as distribution of marking load. During such meetings, the middle leaders (here we mean the head of departments and level heads in schools) will be in charge of regulating the meeting, picking up feedback from teachers, and ensuring that tasks are completed and good practices are in place. With the shift in orientation from scholarly academic to learner-centred orientation, heads of the mathematics department in JCs have the responsibility to remind teachers to tease out the non-examinable components such as developing a positive attitude towards the subject and incorporating engaging and enjoyable *learning experiences* in their classroom activities.

In Section 5.3, we will describe how JC mathematics teachers and heads of department exercise curriculum leadership roles in Context 3, and attempt to show how decisions made in Context 3 affect Context 4.

5.2 *Context 4*

This quadrant focuses on curriculum decisions with a particular scope and immediate implications. Typically, decisions belonging to this context deal with the day-to-day work and activities in the classroom, and often go down to the level of the individual student. Of course, such curriculum decisions are carried out by the classroom teacher. Because each student is unique and has his or her specific needs, the teacher must exercise professional discretion in choosing what to focus on, how and when to teach a certain concept, how and when to elicit response from students and what to do next. Though classroom teachers may feel that they do not have a part to play in making curriculum decisions at the policy level (Context 1), or in drafting the syllabus (Context 2), or a limited and minimal part to play in deciding on the scheme of work (Context 3), they certainly cannot deny that they have a key role in making curriculum decisions in the classroom (Context 4).

Decisions emerging from this context typically address the specific needs pertaining to the specific profile of students. For instance, if a teacher reported that JC1 students experienced difficulties with the manipulation of the trigonometric form of complex numbers, the head of department or the level head might suggest that a revision on trigonometrical functions and their identities may be needed.

5.3 *Exercising curriculum leadership in Contexts 3 and 4*

An exemplar of learning experience created by and for teachers is given in Figure 2. It connects two different topics in the Mathematics 9758 Syllabus.

JC mathematics teachers can exercise their academic knowledge and professional discretion to design suitable learning experience to augment the usual tutorial discussion of 'traditional' examination style questions. The exemplar in Figure 2 provides the content material for deeper exploration and discussion for the topic of linear regression, making use of prior knowledge in vectors. A lesson whose *goal* is to create the stated learning experience can be designed with the stated learning objectives, and must include suitable scaffolding, for example, with the placement of appropriate group *tasks* and *activities* in the form of discussion episodes and worksheets. Crucially, the prior knowledge of vectors in three dimensions is to be invoked with particular emphasis on the dot product of vectors. The subsequent activity guides the students to extend the geometric inequality satisfied by three dimensional vectors to a more general algebraic inequality called the 'Cauchy-Schwarz inequality'. This exemplar shows how JC teachers can be active agents in enacting the H2 Mathematics curriculum by explicating the precise connections between seemingly unrelated topics within the perimeter of the mathematics syllabus. Notice that the choice of the topic involves several topics within the syllabus, e.g., inequalities, quadratic functions, vectors, statistics—an

Regression and Correlation

<u>Lesson Objective</u>. Exploiting inequalities associated to the generalized dot-product (or else known as the Cauchy-Schwarz inequality) we get the students to justify that the linear correlation coefficient, r, has magnitude at most 1.

<u>Problem:</u> Given a set of n data points $\{(x_i, y_i) : i = 1, 2, \ldots, n\}$, the linear correlation coefficient defined by $r = \dfrac{\sum_{i=1}^{n}(x_i - \bar{x})(y_i - \bar{y})}{\sqrt{\sum_{i=1}^{n}(x_i - \bar{x})^2 \sum_{i=1}^{n}(y_i - \bar{y})^2}}$ has the amazing numerical property that $-1 \le r \le 1$. How can one prove that this inequality is true?

<u>Mathematical Pedagogical Content Knowledge</u>. To simplify matters, perhaps we should start by taking $n = 3$. If we take $n = 1$ or 2, the inequality is less interesting. Let $a_i = x_i - \bar{x}$ and $b_i = y_i - \bar{y}$ (for $i = 1, 2, 3$). Proving $|r| \le 1$ is equivalent to showing that for any triplets of real numbers a_i, b_i, $(i = 1, 2, 3)$ one has

$$|a_1 b_1 + a_2 b_2 + a_3 b_3| \le \sqrt{a_1^2 + a_2^2 + a_3^2}\sqrt{b_1^2 + b_2^2 + b_3^2}$$

This looks very familiar…if you at look it through the lens of the dot-product of vectors.

More precisely, let $\mathbf{a} = \begin{pmatrix} a_1 \\ a_2 \\ a_3 \end{pmatrix}$ and $\mathbf{b} = \begin{pmatrix} b_1 \\ b_2 \\ b_3 \end{pmatrix}$, then the above inequality is precisely:

$$|\mathbf{a} \cdot \mathbf{b}| \le |\mathbf{a}||\mathbf{b}|.$$

This inequality is known to be true because of the definition of $\mathbf{a} \cdot \mathbf{b}$ as $|\mathbf{a}||\mathbf{b}| \cos\theta$, where θ is the angle between the vectors $\mathbf{a}$ and $\mathbf{b}$. Deeper thought probes one to ask if this inequality holds independent of this geometrical background. Fortunately, this is the case and can even be proven in full generality for any positive integer n, not just 3. We proceed to do so by relying on nothing but quadratic functions. Given the values $a_1, a_2, \ldots, a_n$ and $b_1, b_2, \ldots, b_n$, we form the following quadratic function $Q(t)$ in t:

$$Q(t) := (a_1 t + b_1)^2 + (a_2 t + b_2)^2 + \cdots + (a_n t + b_n)^2.$$

Because it is the sum of squares of real numbers, it follows that for all t, $Q(t) \ge 0$.

Expanding and arranging, we rewrite $Q(t) = \left(\sum_{i=1}^{n} a_i^2\right)t^2 + \left(2\sum_{i=1}^{n} a_i b_i\right)t + \left(\sum_{i=1}^{n} b_i^2\right)$, where we denote

$$A = \left(\sum_{i=1}^{n} a_i^2\right), B = \left(2\sum_{i=1}^{n} a_i b_i\right), C = \left(\sum_{i=1}^{n} b_i^2\right).$$

Since $Q(t) \ge 0$ holds for all real t and A is positive, it follows that the discriminant of the quadratic $Q(t)$ is at most 0. Thus, $B^2 - 4AC \le 0$ is equivalent to $\left(\sum_{i=1}^{n} a_i b_i\right)^2 \le \left(\sum_{i=1}^{n} a_i^2\right)\left(\sum_{i=1}^{n} b_i^2\right)$. Taking square roots then yield the desired inequality, which is called the Cauchy-Schwarz inequality. Fitting everything back to the original setting, it is immediate that $|r| \le 1$.

<u>Further exploration</u>. The equality in the Cauchy-Schwarz inequality holds if and only if $\mathbf{a}$ is a scalar multiple of $\mathbf{b}$. The students may be invited to prove the aforementioned sharpness condition of the Cauchy-Schwarz inequality, and hence deduce the linear properties that arise when $|r| = 1$.

Figure 2. Exemplar for Learning Experience in Regression and Correlation

illustration of the richness and interconnectedness of mathematics as a discipline at a level that can be appreciated by JC students. Thus, the learning experience creates a natural platform for JC mathematics teachers, together with their heads of department, to be active agents who are responsible for translating intended curriculum to enacted curriculum, and which subsequently will be received by the learners as experienced curriculum. A better understanding of the big picture of the syllabus and curriculum changes, via Eisner's model of contexts of curriculum decision-making, provides the impetus for JC mathematics teachers to passionately take on their role as curriculum leaders, in their own professional capacity. All these will ultimately result in the concretization of goals, tasks and activities in the JC mathematics classroom.

6 Conclusion

Our preceding development traces the major educational initiatives brought about by decisions arising from Context 1 (policy level), and details the direct implications on the 'A' Level Mathematics Syllabus (Context 2) along the timeline. Curriculum decisions emanating from Context 1 certainly have direct repercussions on activities pertaining to the rest of the contexts. This often creates the false impression that MOE makes all the important decisions such that teachers at the school level often see themselves as passive implementers of the syllabus drawn out by MOE. However, the framework of contexts of curriculum decision making indicates that teachers and heads of department in the school have the responsibility to understand the purpose for the changes in syllabi, their shifts in educational orientations, and how these are translated into the scope and sequence of teaching and learning experiences and assessments.

This chapter brings to the awareness of JC mathematics teachers the recent shift of curriculum orientation towards learner-centred approaches, and informs about the professional decisions which need to be considered in terms of the context in which and for which the curriculum is developed for 'A' Level Mathematics. It is hoped that, through this vicennial walk through the Singapore 'A' Level Mathematics, JC mathematics teachers see themselves as active curriculum decision-makers, directly impacting the learners of mathematics in JCs.

Acknowledgement

The first author would like to thank his MME colleagues for sharing their wealth of knowledge on Mathematics curriculum in Singapore during the process of the writing of this book chapter.

References

Bobbitt, F. (2009). Scientific Method in curriculum making. In D. J. Flinders, & S. J. Thornton (Eds.), *The Curriculum Studies Reader* (3rd ed.). New York: Routledge.

Deng, Z., Gopinathan, S., & Lee, C. K. E. (Eds.) (2013*). Globalization and the Singapore Curriculum: From Policy to Classroom.* Singapore: Springer.

Eisner, E. (2002). *The educational imagination: On the design and evaluation of school programs* (3rd ed.). Upper Saddle River, NJ: Pearson Education.

Goh, C. B., & Gopinathan, S. (2008). The development of education in Singapore since 1965. In S. K. Lee, C. B. Goh, B. Fredriksen, & J. P. Tan (Eds), *Toward a Better Future* (Washington DC, The International Bank for Reconstruction and Development / The World Bank).

Guskey, T. R. (2002). Professional Development and Teacher Change. *Teachers and Teaching: theory and practice, 8*(3/4), Cafax Publishing, Taylor and Francis.

Hairon, S., Tan, K., Lin, T.-B., & Lee, M. M. (2017). Grappling with curriculum leadership theory in schools. In K. Tan, M. A. Heng, & C. Ratnam-Lim (Eds.), *Curriculum Leadership by Middle Leaders: Theory, design and practice* (Chapter 2, pp. 10-25). Routledge.

Ho, W. K., Toh, P. C., Teo, K. M., Zhao, D. S., & Hang, K. H. (2018). *Beyond School Mathematics.* In T.L. Toh, E. G. Tay, & B. Kaur (Eds.), *Mathematics Education in Singapore.* Springer Series: Mathematics Education – an Asian Perspective, Springer.

Lim-Ratnam, C., Atencio, M., & Lee, C. K. E. (2016). Managing the paradox of control: the case of ground-up implementation of active learning in Singapore's primary schools. *Educational Research for Policy and Practice, 15*, 231-246.

Malthouse, R., Roffey-Barentsen, J., & Watts, M. (2014). Reflectivity, reflexivity and situated reflective practice. *Professional Development in Education, 40*(4), 597-609.

Ministry of Education. (1997). *Shaping Our Future: Thinking Schools, Learning Nation.* Speech by Prime Minister Goh at the Opening of the 7th International Conference on Thinking on Monday, 2 June 1997, at 9.00am at the Suntec City Convention Centre Ballroom.

Ministry of Education. (2005). Mathematics (2005) (Revised Syllabus 9233). Singapore Examinations and Assessment Board & University of Cambridge Local Examination Syndicate, Ministry of Education, Singapore.

Ministry of Education. (2006). Mathematics Higher 2 (2006) (Syllabus 9740). Singapore Examinations and Assessment Board & University of Cambridge Local Examination Syndicate, Ministry of Education, Singapore.

Ministry of Education. (2017). Mathematics Higher 2 (2017) (Syllabus 9758). Singapore Examinations and Assessment Board & University of Cambridge Local Examination Syndicate, Ministry of Education, Singapore.

Ng, P. T. (2005). Students' perception of change in the Singapore education system. *Educational Research for Policy and Practice, 3*(1), 77-92.

Ratnam-Lim, C. (2017). Decision-making in curriculum leadership. In K. Tan, M. A. Heng, & C. Ratnam-Lim (Eds.), *Curriculum Leadership by Middle Leaders: Theory, design and practice* (Chapter 4, pp. 42-57). Routledge.

Schiro, M. S. (2013). *Curriculum Theory: Conflicting Visions and Enduring Concerns.* Sage Publications Inc., Boston College.

Schwab, J.J. (1973). The Practical 3: Translation into curriculum. *The School Review, 81* (4), 501-522.

Teo, C.H. (2000). Dynamic school leaders and schools- making the best use of autonomy. Speech presented by Teo Chee Hean, Minister for Education and Second Minister for Defence, at Mandarin Hotel, Singapore. Retrieved September 10, 2015 from http://www.nas.gov.sg/archivesonline/speeches/view-html?filename=2000011201.htm

Thomas, P. L. (2014). *Why Teachers' Voices Matter in Education Reform Rebate.* Alternet: https://www.alternet.org/education/why-teachers-voices-matter-education-reform-debate

Thompson, N., & Pascal, J. (2012). Developing critically reflective practice. *Reflective Practice, 13*(2), 311-325.

Tyler, R. W. (1949). *Basic principles of curriculum and instruction.* Chicago: University of Chicago Press.

Whitefield, R. C. (Ed.). (1971). *Disciplines of curriculum.* London: McGraw-Hill.

Chapter 14

Probability: Theory and Teaching

YAP Von Bing

Since the middles ages, probability has fascinated gamblers and philosophers. A satisfactory mathematical foundation was only formulated in the twentieth century by Kolmogorov. What preparation should teachers receive, in order to teach probability in the classroom? This chapter presents personal views, supported by the great textbook of Freedman, Pisani and Purves. The teacher ought to appreciate Kolmogorov's axioms in the case of discrete random variables, to use the frequency interpretation to give meaning to probability statements, and to help students distinguish between situations where the theory applies and those where there is doubt.

1 Introduction

The human struggle against the vagaries of life gave us vivid notions like fate, risk, and uncertainty, which may be precursors to the concept of chance, or probability. The famous correspondence between Pascal and Fermat in 1654 about a gambling dispute laid the foundational principles of probability theory (Apostol, 1969), and paved the way for the rapid development of applications in many practical problems. Kolmogorov's axiomatisation in *Foundations of the Theory of Probability* (1933) is widely considered a crowning achievement of probability theory. The field continues to flourish in response to new problems in human affairs, spanning wide swathes of empirical sciences and technological innovations. Unlike other mathematical fields like algebra and calculus, probability means different things to different intellectual communities: philosophers, mathematicians, engineers, etc. The objective of this chapter is twofold: (a) to outline the Kolmogorov axioms in the case of certain discrete random variables and the connections with several interpretations; (b) to present some ideas for teaching the frequency interpretation to pre-university students, via a set of solved problems and a set of discussed problems. Most of the teaching ideas are included in the classic textbook *Statistics* by Freedman, Pisani, and Purves (2007), or inspired by it. Any teacher of probability will do well to first use chapters 13 to 15 of *Statistics*. After gaining some experience, the teacher may consult the education literature for philosophical or pedagogical underpinnings in order to sharpen the practice. Excellent references include Cobb and Moore (1997) and chapter 7 of the *International Handbook of Mathematics Education* (Bishop, Clements, Keitel-Kreidt, Kilpatrick, & Laborde, 1996).

2 The Theory of Probability

This section is concerned with defining probability on a countable sample space S. Some important examples are:

-The nonnegative integers $Z_+ = \{0, 1, 2, \dots\}$, an infinite set.
-The finite set $\{0, 1, \dots, n\}$, where n is a positive integer.

For this chapter, it is appropriate and useful to view Z_+ as a concrete representative of S, since all discrete-valued random variables can be thought of as supported on Z_+.

2.1 *Kolmogorov's axioms*

Every subset of S is called an event. In particular, the empty set and S are both events. For every event A, let $P(A)$ be a number in [0,1], such that two conditions are satisfied:

(1) $P(S) = 1$.

(2) If $A_1, A_2, \ldots$ are mutually disjoint events with union A, then $P(A) = \sum_{i=1}^{\infty} P(A_i)$.

We call $P(A)$ the probability of A. If $P(B) > 0$, define the conditional probability of A given B as

(3) $P(A \mid B) = \frac{P(A \cap B)}{P(B)}$.

The pair (S, P) is called a probability space. Kolmogorov (1933) showed that a probability P can be defined for much more complicated sample spaces, including uncountable sets, such as the unit interval of real line, $[0,1]$, provided a technical concession is made. The price to pay is that for some subset F of S, $P(F)$ is undefined, i.e., not all subsets of S are events. For example, a uniform random number in $[0,1]$ corresponds to a probability P such that for any subinterval $[a, b]$, $P([a, b]) = b - a$. However, there exist subsets of $[0,1]$ that do not have a defined probability. Such exceptional sets are too complicated to visualise, and take lots of technical work to describe. They are much more complex than, for instance, the union of countably many disjoint intervals. The constraint has no practical consequence on the successful applications of probability to a vast array of real problems. The practitioners can take comfort in Kolmogorov's axioms, like they do with the hardware and machine codes underlying all computer languages.

In describing a probability space, many authors provide some intuition, for instance "The sample space S is the set of all possible outcomes of a random experiment", or "Event A happens if any of the elements in A is obtained from the experiment". Such motivations are very valuable for learning the subject, but a probability space need not correspond to any actual experiment. Kolmogorov's axioms do not explain what $P(A)$ means in other terms; in the language of logic, it is a known as a *primitive*. The axioms merely impose a logically consistent structure on the events. Another way to see this point is that the structure described so far is sufficient to solve examination problems of the following type. Given $P(A \cap B) = 0.2$ and $P(A) = 0.5$, determine the value $P(\text{not } B|A)$. There is no need to inquire what $P(A) = 0.5$ actually means.

Many probability spaces do describe random experiments. Suppose S consists of $s > 1$ distinct real numbers $\{x_1, x_2, ..., x_s\}$, and let $k_1, k_2, ..., k_s$ be positive integers with sum k. Then setting $P(\{x_i\}) = k_i/k$ gives a probability space, as can be readily checked. It describes a random draw from a box of k identical balls, of which k_1 are labelled x_1, k_2 are labelled x_2, etc., called a box model in Freedman et al. (2007). In the same vein, a finite set of positive rational numbers summing to 1 describes a similar experiment. But if some of the numbers are irrational, then it is impossible to make the analogy. The impossibility also applies to an infinite set of positive numbers summing to 1, even if they are all rational, like ½, ¼, ⅛, ...

The Kolmogorov axioms can be used to construct general random variables, discrete or continuous or otherwise, technically as a measurable function from a suitable probability space (S, P) to the real numbers. The technical difficulty, including the concept of measurability, disappears when the axioms are specialised to discrete random variables, as outlined above. Furthermore, if a discrete random variable is finite with rational probabilities, then it can be represented as a box model. This observation suggests that getting students to become comfortable with the box model is important for developing a good understanding of discrete random variables.

2.2 *The meaning of probability*

Children learn addition and multiplication of natural numbers by counting various objects, not by deducing formulae from axioms like commutativity, associativity, etc. Likewise, realistic applications of probability should be made available to a student before encountering the subject axiomatically. This is a difficult task, as can be seen by a comparison with plane geometry. The Greeks' axiomatisation of geometry, most famously narrated by Euclid, was a remarkable accomplishment. Terms like length and angle are abstract, but have persuasive visual meanings, which greatly help students understand the deduction of propositions from premises (axioms or other propositions). In contrast, probability does not admit a dominant meaning, which accounts for the inclusion of "different methods of assigning probability" in most textbooks. We do not see "different methods of assigning angle" in a typical geometry textbook. The common assignment methods are

(a) Equal likelihood: If there are s possible outcomes, the probability of an outcome is $1/s$.

(b) Relative frequency: The probability of an outcome is the limit of the fraction of times the outcome occurs in infinitely many repetitions of the experiment, conducted independently and under identical conditions.

(c) Personal belief: The probability of an outcome is the amount of belief the person has about the propensity of its occurrence.

Method (a) has severe limitations. In many problems, such as the box experiment above, the interesting outcomes have unequal probabilities. It is also not applicable to an experiment with infinitely many outcomes. Such experiments often arise in statistical analyses of real experiments, the majority of which have finitely many outcomes. Method (c) offers a lot of flexibility, mainly because it does not admit any analysis, so can pose serious challenges in communication. Both (a) and (c) do not explain what probability means, unlike (b). A drawback of (b) is that it presupposes that the experiment can be repeated many times, independently and under the same conditions, and that the fractions of interest have a limit. This has implications on applications, which will be elaborated later.

2.3 *The frequency interpretation*

In my opinion, the frequency interpretation, the assignment (b), is the best approach for beginners. It builds naturally on students' prior knowledge of data summaries, dovetails seamlessly with a pedagogy that incorporates data analysis, such as that advocated by Cobb and Moore (1997), and works very well with computer simulation, a powerful aid for understanding the frequency interpretation of probability. There is then no need to contemplate other ways of assigning probability beyond a brief mention.

Suppose an experiment is repeated n times, independently and under the same conditions. Informally, "independent" means the outcome of any trial does not influence that of any other trial. Let $n(A)$ be the number of times event A occurs. The frequency approach says that as n goes to infinity, $n(A)/n$ converges to a number $P(A)$, namely the probability of A. Also, for event B with $P(B) > 0$, define $P(A|B)$ as the limit of $n(A \cap B)/n(B)$, i.e., the fraction of occurrences of B with A co-occurring. We now verify that the frequency definitions satisfy the Kolmogorov axioms.

(1) Clearly, $n(S) = n$ always, so $P(S) = 1$.
(2) Let $A_1, A_2, \dots$ be mutually disjoint events with union A. Then $n(A) = n(A_1) + n(A_2) + \cdots$, and it follows that $P(A) = \sum_{i=1}^{\infty} P(A_i)$.
(3) As n goes to infinity, the fraction $n(A \cap B)/n(B) = [n(A \cap B)/n]/[n(B)/n]$ converges to $P(A \cap B)/P(B)$.

Thus we have shown that (1), (2) and (3) follow from the frequency approach. In other words, these statements have become theorems. In contrast, personal belief is weaker than Kolmogorov's axioms. For instance, it is not clear why the belief in the disjoint union $A \cup B$ should be equal to the sum of the beliefs in A and in B. Indeed, there is convincing evidence that human intuition can attach less belief to an event than its subset (Tversky & Kahneman, 1982).

If the experiment can be repeated independently under the same conditions, the observed fraction $n(A)/n$ is an estimate of $P(A)$, and it gets better as n gets large. Note that it is generally not true, though possible, that $n(A)/n$ happens to equal $P(A)$. But one would not be able to tell when this is the case.

Mathematically, the frequency interpretation as described above is informal. For instance, it is not clear what "converges" means. This is not an impediment to an intuitive grasp of the meaning of probability. For a rigorous treatment, the reader may refer to Kerrich (1946) for a very readable account, or the work of von Mises in 1919 (von Mises 1981) which later gave way to Kolmogorov's system.

3 The Teaching of Probability

This section contains a menu of problems and solutions which aim to strengthen the appreciation of the frequency interpretation of probability. Most are positive examples, where probability is often relevant. Also included are some negative examples, where probability, at least the frequency interpretation, seems irrelevant. The demarcation of the limit of applications should be more widely practised in general mathematics instruction. Many more exercises can be found in Freedman et al. (2007). For some activities used in pedagogical research, refer to the *International Handbook of Mathematics Education* (1996). The examples may guide the teacher to design problems in other contexts.

3.1 *Basic Problems*

To solve all the following problems, the student needs explicit instruction on the frequency interpretation of probability, the addition rule for the special case of mutually exclusive events or the general case:

$$P(A \cup B) = P(A) + P(B) - P(A \cap B).$$

The multiplication for the special case of independent events or preferably the general case:

$$P(A \cap B) = P(A|B)\, P(B) = P(B|A)\, P(A).$$

Problem 1. When a coin is tossed, one observes either head (H) or tail (T). For a fair coin, the two outcomes are equally likely: 0.5. What does $P(\text{H}) = 0.5$ mean?

Solution. Toss the coin n times independently under the same conditions. Let $n(H)$ be the number of times head is obtained. As n goes to infinity, $n(H)/n$ goes to 0.5. Kerrich (1946) was based on such an experiment carried out while he was a prisoner-of-war.

Problem 2. When a fair die is rolled, the number of spots observed is equally likely to be 1, 2, 3, 4, 5 or 6. The experiment is repeated 180 times independently and under the same conditions. Roughly how many times will the number of spots be divisible by 3?

Solution. Each outcome has probability 1/6. The event is $\{3, 6\}$, so by the addition rule, the probability is $1/6 + 1/6 = 1/3$. This will be observed around $1/3 \times 180 = 60$ times.

Problem 3. A box contains 8 identical balls; 5 are white and 3 are black. In a random draw, each ball is equally likely to be chosen. Two random draws are made without replacement, i.e., after the first ball is drawn, another random draw is made from the remaining 7 balls. Define the events:

$W_1 = \{1^{st}$ draw gets a white ball$\}$,
$W_2 = \{2^{nd}$ draw gets a white ball$\}$.

What are the values of $P(W_1)$, $P(W_2|W_1)$, $P(W_1 \cap W_2)$, $P(W_2)$, and $P(W_1|W_2)$?

Solution. Since each ball is equally likely, $P(W_1) = 5/8$. After a white ball has been drawn, there are 4 white and 3 black balls left, so $P(W_2|W_1) = 4/7$. By the multiplication rule,

$$P(W_1 \cap W_2) = P(W_1)\, P(W_2|W_1) = 5/8 \times 4/7 = 5/14.$$

Define B_1 and B_2 similarly with respect to black balls. There are two different ways for W_2 to occur, depending on the first draw, so by the addition and multiplication rules,

$$P(W_2) = P(W_1 \cap W_2) + P(B_1 \cap W_2) = 5/8 \times 4/7 + 3/8 \times 5/7 = 5/8.$$

Finally, $P(W_1|W_2) = P(W_1 \cap W_2) / P(W_2) = 4/7$, by the multiplication rule. Students are often surprised that $P(W_2) = P(W_1)$. The mind may find it hard to contemplate the conditional probability $P(W_1|W_2)$, which seems to reverse the experimental sequence. The next two problems can help overcome these cognitive barriers.

Problem 4. Suppose the outcomes of 5,600 repeated experiments of Problem 3 are recorded in a table of 5,600 rows and 2 columns. For instance, if the first trial yields a white and then a black ball, the first row has (W, B); if the second trial yields a black and then a white ball, the second row has (B, W). Thus, each row must be either (W, W), (W, B), (B, W) or (B, B). Roughly how many rows have

(i) W in the first column?
(ii) W in the second column?
(iii) (W, W)?
(iv) Among the rows in which the second column is W, roughly how many of them have W in the first column?

Solution.

(i) This concerns $P(W_1)$, which is 5/8. There are around
$5,600 \times 5/8 = 3,500$ W's in the first column.

(ii) Since $P(W_2) = 5/8$, there are around $5,600 \times 5/8 = 3,500$ W's in the second column.

(iii) $P(W_1 \cap W_2) = 5/8 \times 4/7$. Thus, there are around
$5,600 \times 5/8 \times 4/7 = 2,000$ rows of (W, W).

(iv) $P(W_1|W_2) = 4/7$. There are around $5,600 \times 4/7 = 3,200$ rows with the required property. Illustrating this probability using frequencies helps the mind ease off the urge to perceive it causally.

Problem 5. Table 1 below is generated from 1,000 simulations of the experiment in Problem 3, using a computer software. Verify the figures are close to the theoretical predictions.

Table 1

Results of 1000 simulations of experiment in Problem 3

First draw	Second draw	Count
W	W	353
W	B	281
B	W	266
B	B	100

Solution. We expect to see (W, W) $1,000 \times 5/8 \times 4/7 \approx 357$ times, to the nearest integer. The actual count is off by about $353 - 357 = -4$. Similarly, the expected count of (W, B) is around $1,000 \times 5/8 \times 3/7 \approx 268$, so the actual count is off by about $281 - 268 = 13$. The discrepancy for (B, W) is $266 - 1,000 \times 3/8 \times 5/7 \approx -2$, and for (B, B), we have $100 - 1,000 \times 3/8 \times 2/7 \approx -7$. Without calculating the standard deviation (SD) of the counts, it is hard to say if the discrepancies are serious. This problem should be revisited when the binomial distribution is taught. For example, the SD for the number of times (W, W) is obtained is the square-root of $1,000 \times 5/14 \times (1 - 5/14)$, which is approximately 15.2. So the discrepancy -4 is merely 26% of the SD.

The randomness in the experiment ought to be emphasised by generating another table, which will look different, and going through it as shown above.

Problem 6. Two draws are made at random from a box of white balls and black balls. The experiment is repeated several thousand times, resulting in a table such as described in Problem 4. Which comparison can help decide whether the draws are made with or without replacement?

  (A) Fraction of W's in column 1 vs fraction of W's in column 2.
  (B) Fraction of W's in column 2 among rows with column 1 showing W, vs fraction of W's in column 2.
  (C) None of the above.

Solution. The fraction of W's in column 1 is approximately the chance of getting a white ball in the first draw, $P(W_1)$. Similarly, the fraction of W's in column 2 is approximately $P(W_2)$. Since these chances are equal in both situations, (A) does not help. The fraction of W's in column 2 among rows with column 1 showing W is approximately $P(W_2|W_1)$, which is equal to $P(W_2)$ if the draws are made with replacement, but is different if the draws are made without replacement. Choose (B).

The data in Problem 5 can be used to make the question more concrete. There, the fraction of W's in column 2 among rows with column 1 showing W is $353 / (353+281) \approx 0.557$, whereas the fraction of W's in column 1 is $(353+281) / 1000 = 0.634$. Putting aside concerns about random fluctuations, one may conclude that the data indicate the draws have been made without replacement.

Problem 7. A black box contains 10 identical balls, some white and some black, but you do not know how many of each there are. You are allowed to make two random draws with replacement, and to repeat this experiment 1,000 times. Table 2 shows the counts you obtained. How many white balls are there in the box?

Table 2

Results of 1000 repetitions of experiment in Problem 7

First draw	Second draw	Count
W	W	376
W	B	238
B	W	245
B	B	141

Solution. If the numbers of white and black balls are equal, we expect the counts to be similar. Instead they suggest that there are more than 5 white balls. Suppose there are 6 white balls and 4 black balls. Then the expected counts are

(W, W): $1,000 \times 0.6 \times 0.6 = 360$,
(W, B): $1,000 \times 0.6 \times 0.4 = 240$,
(B, W): $1,000 \times 0.4 \times 0.6 = 240$,
(B, B): $1,000 \times 0.4 \times 0.4 = 160$.

So this hypothesis seems likely. If there are 7 white balls and 3 black balls, we expect $1,000 \times 0.7 \times 0.7 = 490$ (W, W)'s, which is rather different from 376, and it gets worse with even more white balls. We can conclude that there are 6 white balls and 4 black balls.

This problem is a good motivation for inferential statistics, including estimation and hypothesis testing. It also shows that some kind of random process is always assumed in statistical inference, a fact which does not get sufficient attention in many textbooks and applications.

The remaining problems deal with independent events. Let A and B be events with positive probabilities. They are said to be independent if any of the following is true:

(i) $P(A|B) = P(A)$;
(ii) $P(B|A) = P(B)$;
(iii) $P(A \cap B) = P(A)P(B)$.

It is straightforward to verify that (i), (ii) and (iii) are equivalent. (iii) can be applied to the case where $P(A)$ or $P(B)$ equals 0, but there is little practical advantage in enlarging the concept of independence this way.

Problem 8. Suppose two coins are tossed, the first one by Alice, and the second one by Bob. Assume that the actions of Alice and Bob have nothing to do with each other. Let H_1 and H_2 be the events that Alice and Bob observed heads respectively. Show that the events are independent.

Solution. Suppose Alice tosses first. By assumption, if Alice has obtained heads, then Bob's coin still behaves like normal. Hence

$P(H_2|H_1)$ must be equal to $P(H_2)$. By definition, H_1 and H_2 are independent events.

We may describe the assumption about the two coins as "mechanistic independence". Roughly speaking, mechanistic independence implies probabilistic independence. It is worthwhile to emphasise the reasoning: assuming mechanistic independence allows us to derive $P(H_2|H_1) = P(H_2)$, which implies probabilistic independence. We do not jump straight from mechanistic independence to probabilistic independence as if it is obvious.

Problem 9. Consider drawing twice at random from a box with some white balls and some black balls. Let W_1 and W_2 be the events of getting a white ball on the first and second draw respectively. Are W_1 and W_2 independent, if the draws are made with replacement? How about without replacement?

Solution. First case: with replacement. If a white ball is drawn first, it is returned to the box before the second draw. Hence $P(W_2|W_1)$ is the fraction of white balls in the box, equal to $P(W_2)$. By definition, W_1 and W_2 are independent.

Second case: without replacement. Now by Problem 3, $P(W_2|W_1)$ and $P(W_2)$ are different, so the events are dependent.

In both Problems 8 and 9, a conditional probability is compared with an unconditional probability, in order to decide whether the events are independent. The notion of probabilistic independence is not as intuitive as mechanistic independence.

Problem 10. Shuffle a deck of playing cards, then deal two, i.e., two cards are drawn at random without replacement. Let

   $A = \{\text{first card is a heart}\}$,
   $B = \{\text{second card is an ace}\}$.

Are A and B independent?

Solution. To answer the question, we need to compute some probabilities. In this case, we compare the product $P(A)P(B)$ with $P(A \cap B)$.

$$P(A)P(B) = 13/52 \times 4/52 = 1/52.$$

$A \cap B$ is the disjoint union of two events, with probabilities given by the multiplication rule

{first is ace of heart, second is ace}: $1/52 \times 3/51$.
{first is non-ace of heart, second is ace}: $12/52 \times 4/51$.

By the addition rule, $P(A \cap B) = 1/52 = P(A) P(B)$. So A and B are independent events. This might seem surprising, since mechanistic independence looks doubtful.

3.2 *Additional explorations*

This subsection presents three topics which are more suitable for class discussion than a written assignment or test. Although convincing resolutions may not be available, they highlight some fundamental issues of interpreting probability in real-life problems.

Global temperature. The top graph in Figure 1 plots the global July temperature for the past 50 years. Do the data look like they are generated by randomly drawing numbers with replacement from a set of numbers?

In the bottom graph, the y-values are generated by randomly drawing with replacement from the original temperatures, using a computer. In contrast to the top graph, there is no clear trend in the bottom graph. We conclude that the top graph does not look like coming from a random process. The Earth is steadily getting warmer over the last 50 years. Many scientists believe the trend is mostly caused by human activities.

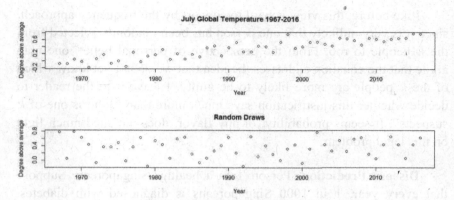

Figure 1. July global temperature (1967-2016) and a randomly generated graph

This is a cautionary example against viewing every data set as coming from a simple random process. To fit the temperature data, a more sophisticated probabilistic model is needed.

Crime Evidence. An old lady was robbed in broad daylight. She could not see the culprit's face, but knew it was a young man with a tattooed right arm. John, who has a tattooed right arm, is a suspect. What does P(John is guilty) mean?

John is either guilty or not guilty. Since the crime is a unique event, it does not seem reasonable to think of his alleged role in the crime as the outcome of a random experiment. Let us take an elementary approach. Suppose there are k people with tattooed right arm in the population of size n. The smaller k is, the lower the likelihood of John being innocent. After all, if there is only one such person, namely John, then he is the culprit. This intuition might lead us to equate P(John is innocent) $= 1 - P$(John is guilty) to the fraction of people with tattooed right arm in the population: k/n, or $(k-1)/n$, in order for the value to agree with intuition in the case $k = 1$. However, people with no tattoo on their right arms are irrelevant here. Perhaps it is more persuasive to say

$$P(\text{John is guilty}) = 1/k.$$

Like before, this view is hard to support by the frequency approach, since it is very unlikely that one person has been randomly selected from the k people to rob. From the perspective of personal belief, one may argue that the equation expresses the idea that we do not yet know which of the k people are more likely to be guilty. I leave it to the reader to decide whether this justification says much more than "John is one of k suspects". It seems probability, of any flavor, does not shed much light on this legal problem.

Disease Prediction. Person T is a healthy Singaporean. Suppose that every year, 1 in 1000 Singaporeans is diagnosed with diabetes. Assuming the future behaves more or less like the present, does it follow that

P(T will be diagnosed with diabetes in 2018) = 1/1000?

Since T is a unique individual, the event {T will be diagnosed with diabetes} is hard to understand from the frequency perspective. Strictly speaking, one will need to clone many copies of T, let them live independently and under the same conditions, then after some years determine how many of them are diabetic. Even if cloning is technically doable, there are many reasons not to do so. So the probability of the event will be practically unknowable. The personal belief approach may then be employed, and it may seem reasonable to use 1/1000 as a rough approximation. Disease risk depends on many factors. If T is over 60 years old, then the probability should be higher. But this process does not seem to end tidily. If T is male, then one has to revise the number, probably upwards. And if he is ethnically Chinese, it has to come down again. It is hard to see whether the successive revisions will stabilise to a fixed number when all the important factors have been taken in account, assuming we even know them all. It seems like the concept of a disease probability for a specific individual is quite elusive, if not meaningless. For a well-defined population of individuals, the prospect appears less daunting, although in this case, I prefer to think about fractions, and to reserve "probabilities" for the occasion when random sampling is applied to the population.

4 Conclusion

This chapter attempts to outline some connections between the mathematical theory, the meaning of probability, and its teaching. It is highly recommended that the box model and the frequency approach be used extensively in the instruction of probability in schools. It is hoped that the problems and solutions convince the teacher that there is indeed much to teach within the seemingly narrow scope, and that they adequately illustrate both the power and limitations of the frequency approach in solving real problems.

References

Apostol, T. M. (1969). *Calculus*. Volume 2 (2nd ed). Wiley.

Cobb, G. W., & Moore, D. S. (1997). Mathematics, Statistics and Teaching. *The American Mathematical Monthly, 104*, 801-823.

Freedman, D. A., Pisani, R., & Purves, R. (2007). *Statistics* (4th ed). Norton.

Bishop, A. J., Clements, K., Keitel-Kreidt, C., Kilpatrick, J., & Laborde, C. (Eds.) (1996). *International Handbook of Mathematics Education*. Netherlands: Kluwer.

Kerrich, J. E. (1946). *An experimental introduction to the Theory of Probability*. Munksgaard.

Kolmogorov, A. N. (1933). *Grundbegriffe der Wahrscheinlichkeitsrechnung*. Berlin: Julius Springer. Translation: *Foundations of the Theory of Probability* (2nd ed) (1956). New York: Chelsea.

Tversky, A., & Kahneman, D. (1982). Judgments of and by representativeness. In D Kahneman, D., Slovic, P., & Tversky, A. (Eds.) *Judgment under uncertainty: Heuristics and biases*. Cambridge, UK: Cambridge University Press.

von Mises, R. (1981) *Probability, Statistics, and Truth*. Dover.

Contributing Authors

CHAN Chun Ming Eric holds a PhD in Mathematics Education and is a lecturer with the Mathematics and Mathematics Education Academic Group at the National Institute of Education, Singapore. He teaches mathematics education courses at the pre-service and in-service (primary) programmes. His research interests include children's mathematical problem solving and mathematical modelling using model-eliciting activities.

CHENG Lu Pien is a Lecturer in the Mathematics and Mathematics Education Academic Group at the National Institute of Education, Nanyang Technological University of Singapore. She received her PhD in Mathematics Education from the University of Georgia (U.S.) in 2006. She specializes in mathematics education courses for primary school teachers. Her research interests include the professional development of primary school mathematics teachers and children's thinking in the mathematics classrooms.

CHEW Chong Kiat is a secondary school mathematics teacher. Presently he is a Lead Teacher and has been teaching for the last 23 years. He has a keen interest in fostering mathematical reasoning and communication and in developing teaching resources for and teaching low progress learners. He is involved in the professional development of Mathematics Teachers as a core team member of the Secondary Mathematics Chapter of the Academy of Singapore Teachers (AST). For his contributions, he was conferred the Associate of the AST in 2013, 2015 and 2017. He is the recipient of the Excellence in Mathematics Teaching Award in 2017. This award recognizes excellent teachers of mathematics in Singapore schools.

CHOY Ban Heng, a recipient of the NIE Overseas Graduate Scholarship in 2011, is currently an Assistant Professor in Mathematics Education at the National Institute of Education. Prior to joining NIE, he taught secondary school students Mathematics for more than ten years, and held the position of Head of Department (Special Projects) before he joined the CPDD in MOE as a Curriculum Policy Officer. He holds a PhD in Mathematics Education from the University of Auckland, New Zealand. Specialising in mathematics teacher noticing, he developed a theory for describing and analysing what mathematics teachers see, and how they think, when they make productive instructional decisions that enhance students' mathematical reasoning. Ban Heng was awarded the Early Career Award during the 2013 MERGA Conference in Melbourne for his excellence in writing and presenting a piece of mathematics education research.

CHUA Boon Liang is an Assistant Professor in mathematics education at the National Institute of Education, Nanyang Technological University in Singapore. He holds a PhD in Mathematics Education from the Institute of Education, University College London, UK. His research interests cover pattern generalisation, mathematical reasoning and justification, and task design. Given his experience as a classroom teacher, head of department and teacher educator, he seeks to help mathematics teachers create a supportive learning environment that promotes understanding and inspire their students to appreciate the beauty and power of mathematics. With his belief that students' attitudes towards mathematics are shaped by their learning experiences, he hopes to share his passion of teaching mathematics with the teachers so that they make not only their teaching more interesting but also learning mathematics an exciting and enjoyable process for their students. He feels honoured to have been awarded Excellence in Teaching by the National Institute of Education in 2009 and 2013.

DINDYAL Jaguthsing is an Associate Professor in the Mathematics & Mathematics Education Academic Group at the National Institute of Education, Nanyang Technological University in Singapore. He teaches mathematics education courses to both pre-service and in-service teachers. He currently has specific interest in teacher noticing and teachers' use of examples in the teaching of mathematics. His other interests include the teaching and learning of geometry and algebra, lesson study and students' reasoning in mathematics specifically related to their errors and misconceptions.

Keiko HINO is Professor of Mathematics Education at Utsunomiya University in Japan. She received her M.Ed. from Tsukuba University and Ph.D. in Education from Southern Illinois University. She began her career as a Research Assistant at Tsukuba University in 1995. After the career as Associate Professor of Mathematics Education at Nara University of Education, she is now working at Utsunomiya University. Dr. Hino has been Professor since 2010 at Utsunomiya University. Her major scholarly interests are students' development of proportional reasoning and functional thinking through classroom teaching, international comparative study on teaching and learning mathematics, and mathematics teachers' professional development. She has authored or co- authored 3 books, publish 25 book chapters, and over 40 journal articles and presented at over 40 conferences, including the International Congress on Mathematics Education, the International Conference of Psychology of Mathematics Education, the East Asia Regional Conference on Mathematics Education, and annual meetings of the Japan Society of Mathematics Education and of Japan Society for Science Education. She is also involved in activities for improving mathematics education as an editor of Japanese Primary and Lower Secondary School Mathematics Textbooks and External-expert for Lesson Study in Mathematics.

HO Weng Kin received his Ph.D. in Computer Science from The University of Birmingham (UK) in 2006. His doctoral thesis proposed an operational domain theory for sequential functional programming languages. He specializes in programming language semantics and is dedicated to the study of hybrid semantics and their applications in computing. Apart from theoretic computer science, his areas of research interest also cover tertiary mathematics education, flipped classroom pedagogy, problem solving, and computational thinking.

Chunlian JIANG is currently an assistant professor in the Faculty of education, University of Macau. She received her M.Ed. from Central China Normal University, China and her PhD from National Institute of Education, Nanyang Technological University, Singapore. Her research interests include mathematical problem solving and problem posing, use of IT in mathematics teaching and learning, and mathematics Olympiad. She has taught mathematics education courses from kindergarten to graduate levels for both pre-service and in-service teachers.

Berinderjeet KAUR is a Professor of Mathematics Education at the National Institute of Education in Singapore. Her primary research interests are in the area of classroom pedagogy of mathematics teachers and comparative studies in mathematics education. She has been involved in numerous international studies of Mathematics Education and was the Mathematics Consultant to TIMSS 2011. She is also a member of the MEG (Mathematics Expert Group) for PISA 2015. As the President of the Association of Mathematics Educators (AME) from 2004-2010, she has also been actively involved in the professional development of mathematics teachers in Singapore and is the founding chairperson of Mathematics Teachers' Conferences that started in 2005 and the founding editor of the AME Yearbook series that started in 2009. She was awarded the Public Administration Medal by the President of Singapore in 2006.

Barry KISSANE is an Emeritus Associate Professor in Mathematics Education at Murdoch University in Perth, Western Australia. From 1985 until his recent retirement, he taught primary and secondary mathematics teacher education students at Murdoch University, except for a period for which he was Dean of the School of Education and an earlier period working and studying at the University of Chicago. His research interests in mathematics education include the use of technology for teaching and learning mathematics and statistics, numeracy, curriculum development, popular mathematics and teacher education. He has written several books and many papers related to the use of calculators in school mathematics, and published papers on other topics, including the use of the Internet and the development of numeracy. In recent years, Barry has been actively involved with professional development of mathematics teachers using calculators for teaching and learning mathematics, both in Australia and in several Southeast Asian countries. Barry has served terms as President of the Mathematical Association of Western Australia (MAWA) and as President of the Australian Association of Mathematics Teachers (AAMT). He has been a member of editorial panels of various Australian journals for mathematics teachers for around 30 years, including several years as Editor of The Australian Mathematics Teacher. A regular contributor to conferences for mathematics teachers throughout Australasia and elsewhere, he is an Honorary Life member of both the AAMT and the MAWA.

LIM Kam Ming is the Deputy Divisional Director of the Office of Academic Administration and Services at the National Institute of Education (NIE), Nanyang Technological University. He is also an Associate Professor with the Psychological Studies Academic Group at NIE. He was conferred the Public Administration Medal (Bronze) by the President of the Republic of Singapore in 2015. He is currently the President of the Educational Research Association of Singapore, Board member of the Asia-Pacific Educational Research Association, and Council member of the World Education Research Association (2015-2019).

LIM Lee Hean is an Associate Professor attached to the Policy and Leadership Studies Academic Group at the National Institute of Education, Nanyang Technological University, Singapore. She held positions as school vice-principal and head of mathematics department prior to receiving an NTU scholarship to pursue her interests in the theory and practice of mentoring, leadership and management. She has been involved in postgraduate and in-service curriculum conceptualisation, design and delivery of courses for professional development.

LIU Huanjia Tracy is a teacher at the Convent of the Holy Infant Jesus (Kellock), Singapore. She holds a PGDE (Pri) from National Institute of

Education and a BBUS (Applied Econs) from Nanyang Technological University. She has taught in the school for seven years. Her interests are in assessment designs in curriculum, and how students learn through self-regulated reflection.

NG Ee Noch is the Head of Department for Information and Communication Technology at a local primary school. He has been implementing flipped learning with his students since 2013 and is currently exploring the concept of flipped mastery as a way to ensure students' mastery of concepts through the flipped learning methodology. He also has a keen interest in using technology with his students to deepen, enhance and excite them in their learning.

NG Swee Fong is an Associate Professor at the National Institute of Education, Nanyang Technological University, Singapore. She holds a master and a PhD in mathematics education, both from the University of Birmingham, United Kingdom. Her general interests include how teacher's pedagogical content knowledge influence the nature of questions and the choice of examples used to support the teaching and learning of mathematics across the curriculum. The teaching and learning of algebra is her special interest.

Christina RATNAM-LIM Tong Li received her PhD for her thesis on Teachers' Beliefs about Literacy Development in Early Childhood Education from Macquarie University (Sydney). She started her professional journey as a teacher working with students in gifted education in a secondary school in Singapore, and then served in the Ministry of Education in various capacities including designing the English Literature Syllabus, producing educational video material, and designing and implementing a curriculum framework for the early childhood education sector in Singapore. Besides teaching about curriculum design and implementation, her research interests include exploring the curriculum affordances of the growth mindset, implementation concerns and issues, and teacher learning and professional development.

TAN Bee Kian Jasmine Susie is a mathematics teacher and year head at a local primary school. She has been teaching mathematics at the Primary School for thirty years. Her interests lie in understanding how children learn and developing effective pedagogy to enhance teaching and learning.

TOH Pee Choon received his PhD from the National University of Singapore in 2007. He is currently an Assistant Professor at the National Institute of Education, Nanyang Technological University. A number theorist by training, he continues to research in both Mathematics and Mathematics Education. His research interests in Mathematics Education include problem solving, proof and reasoning, and the teaching of mathematics at the undergraduate level.

TOH Tin Lam is an Associate Professor and currently the Deputy Head of the Mathematics and Mathematics Education Academic Group in the National Institute of Education, Nanyang Technological University, Singapore. He obtained his PhD from the National University of Singapore in 2001. He continues to do research in mathematics as well as mathematics education. He has published papers in international scientific journals in both areas.

VAPUMARICAN Rashidah is a Lead Teacher (Mathematics) at the Convent of the Holy Infant Jesus (Kellock), Singapore. She holds a Master in Education in Curriculum and Teaching and has taught in a primary school for more than 20 years. Her interests are in expanding children's knowledge in problem solving and assessment design in Mathematics.

WONG Lai Fong has been a mathematics teacher for over 20 years and is known for her efforts in engaging students with fresh and creative strategies in the study of Mathematics. For her exemplary teaching and conduct she was given the President's Award for Teachers in 2009. She sets the tone for teaching the subject in her school as a Head of Department (Mathematics) from 2001 to 2009, a Senior Teacher and then a Lead Teacher for Mathematics subsequently. She was awarded a

Post-graduate Scholarship by the Singapore Ministry of Education to pursue a Master of Education in Mathematics which she has completed in 2014. Currently, she is involved in several Networked Learning Communities looking at ways to infuse mathematical reasoning, metacognitive strategies, and real-life context in the teaching of mathematics. Lai Fong is active in the professional development of mathematics teachers and in recognition of her significant contribution toward the professional development of Singapore teachers, she was conferred the Associate of Academy of Singapore Teachers in 2015 and 2016. She is currently seconded as a Teaching Fellow in the National Institute of Education, and is also an executive committee member of the Association of Mathematics Educators.

YAP Von Bing obtained a B.Sc. (Hons) in Mathematics and a M.Sc. in Applied Mathematics, both from the National University of Singapore (NUS), and then a Ph.D. in Statistics from the University of California. Since 2004, he has been teaching at the Department of Statistics and Applied Probability in NUS, where he is an Associate Professor. His main interest is applied statistics, the interface between mathematical theory and science. Other fields that fascinate him include probability, molecular evolution, comparative genomics, ecology, and the philosophy and practice of teaching abstract concepts.

Joseph Boon Wooi YEO is a Lecturer in the Mathematics and Mathematics Education Academic Group at the National Institute of Education, Nanyang Technological University, Singapore. He is the first author of the New Syllabus Mathematics textbooks used in many secondary schools in Singapore. His research interests are on innovative pedagogies that engage the minds and hearts of mathematics learners. These include the use of an inquiry approach to learning mathematics (e.g. guided-discovery learning and investigation), ICT, and motivation strategies to arouse students' interest in mathematics (e.g. catchy maths songs, amusing maths videos, witty comics, intriguing puzzles and games, and real-life examples and applications). He is also the Chairman of Singapore and Asian Schools Math Olympiad (SASMO) Advisory

Council, and the creator of Cheryl's birthday puzzle that went viral in 2015.

Joseph Kai Kow YEO is a Senior Lecturer in the Mathematics and Mathematics Education Academic Group at the National Institute of Education, Nanyang Technological University, Singapore. Before joining the National Institute of Education in 2000, he held the post of Vice Principal and Head of Mathematics Department in secondary schools. As a mathematics educator, he teaches pre- and in-service as well as postgraduate courses in mathematics education and supervises postgraduate students pursuing Masters degrees. His publication and research interests include mathematical problem solving at the primary and secondary levels, mathematics pedagogical content knowledge of teachers, mathematics teaching in primary schools and mathematics anxiety.